Lecture Notes in Mathematics

Edited by A. Dold and B. Eckmann

411

Hypergraph Seminar

Ohio State University 1972

Edited by Claude Berge and Dijen Ray-Chaudhuri

Springer-Verlag
Berlin · Heidelberg · New York 1974

Claude Berge
Université de Paris VI
Laboratoire de Calcul des Probabilites
9, Quai St. Bernard – Tour 56
Paris 5e./France

Dijen Ray-Chaudhuri
Ohio State University
Dept. of Mathematics
Colombus Ohio 43210/USA

Library of Congress Cataloging in Publication Data

Working Seminar on Hypergraphs, Ohio State University, 1972.
Hypergraph seminar.

(Lecture notes in mathematics, 411)
Bibliography: p.
1. Hypergraphs--Congresses. I. Berge, Claude, ed. II. Ray-Chaudhuri, Dijen, 1933- ed. III. Title. IV. Series.
QA166.W67 1972 511'.5 74-17434

AMS Subject Classifications (1970): 05-02, 05B25, 05C15

ISBN 3-540-06846-5 Springer-Verlag Berlin · Heidelberg · New York
ISBN 0-387-06846-5 Springer-Verlag New York · Heidelberg · Berlin

Offsetdruck: Julius Beltz, Hemsbach/Bergstr.

FOREWORD

This volume constitutes the proceedings of the first Working Seminar on Hypergraphs, held at the Ohio State University in Columbus (Ohio), from August 16 to September 9, 1972. The Department of Mathematics invited a large number of combinatorialists (from Canada, France, Iceland, Hungary and U.S.A.) to present and discuss results related to the new area of Hypergraph Theory. Since Hypergraphs can be viewed as generalizations of Graphs, or of Matroids, or of Finite Geometries, the main task was to unify concepts and terminology.

We especially thank Professor Arnold E. Ross, Chairman of the Department of Mathematics at OSU, for his encouragement and counsel before and during the conference. His enthusiasm for this new field of research and the constant support he gave to the young mathematicians were a great source of inspiration for all of us. We would also like to thank U.S.R. Murty, R.P. Gupta, K.I. Chang, N.R. Grabois and Alan P. Sprague for their help in editing this volume.

Claude BERGE   Dijen K. RAY-CHAUDHURI

TABLE OF CONTENTS

PART I : GENERAL HYPERGRAPHS

C. BERGE

Isomorphism Problems for Hypergraphs 1

C. BERGE

Nombres de coloration de l'hypergraphe h-parti complet . . 13

C. BERGE and M. SIMONOVITS

The Coloring Numbers of the Direct Product of Two Hypergraphs 21

J.C. BERMOND

Graphe représentatif de l'hypergraphe h-parti complet . . 34

R. BONNET and P. ERDÖS

The Chromatic Index of an Infinite Complete Hypergraph : a Partition Theorem 54

V. CHVÁTAL

Intersecting Families of Edges in Hypergraphs Having the Hereditary Property 61

W.H. CUNNINGHAM

On theorems of Berge and Fournier 67

P. ERDÖS

Extremal Problems on Graphs and Hypergraphs 75

V. FABER

Hypergraph Reconstruction 85

J.C. FOURNIER

Une condition pour qu'un hypergraphe, ou son complémentaire, soit fortement isomorphe à un hypergraphe complet 95

P. HANSEN and M. LAS VERGNAS

On a Property of Hypergraphs with no cycles of Length Greater than 2 99

M. LAS VERGNAS

Sur les hypergraphes bichromatiques 102

L. LOVÁSZ

Minimax Theorems for Hypergraphs 111

J.C. MEYER

Quelques problèmes concernant les cliques des hypergraphes h-complets et q-parti h-complets 127

R. RADO

Reconstruction Theorems for Infinite Hypergraphs 140

M. SIMONOVITS

Note on a Hypergraph Extremal Problem 147

F. STERBOUL

Sur une conjecture de V. Chvátal 152

F. STERBOUL

On the Chromatic Number of the Direct Product of Hypergraphs 165

PART II : GRAPHS, MATROIDS, DESIGNS

V. CHVÁTAL and L. LOVÁSZ

Every Directed Graph Has a Semi-kernel 175

T.A. DOWLING and D.G. KELLY

Elementary Strong Maps and Transversal Geometries 176

P. ERDÖS

Some Problems in Graph Theory 187

T. HELGASON

Aspects of the Theory of Hypermatroids 191

W. PULLEYBLANK and Jack EDMONDS

Facets of 1-Matching Polyhedra 214

W.T. TUTTE

Chromials 243

R.M. WILSON

Some Partitions of All Triples into Steiner Triple Systems 267

PART III

UNSOLVED PROBLEMS 278

LIST OF SYMBOLS

$H = (X,\mathcal{E}) = (E_i / i \in I)$ = hypergraph with vertex set $X = \bigcup E_i$

K_n^h	complete h-uniform hypergraph with n vertices
$L(H)$	linegraph of H : simple graph representing the edges of H
$R(H)$	representing multigraph of H
H^*	dual of H
H_A	subhypergraph of H induced by $A \subseteq X$
$(E_i / i \in J)$	partial hypergraph of H generated by $J \subseteq I$
$H \times A$	section hypergraph of H by $A \subseteq X$
$n(H)$	order of H
$r(H)$	rank of H
$\chi(H)$	weak chromatic number : no edge with more than one element is monochromatic
$\gamma(H)$	strong chromatic number : no two vertices in the same edge have the same color
$\beta(H)$	weak stability number : maximum cardinality of a set $S \subseteq X$ containing no edge with more than one element
$\alpha(H)$	strong stability number : maximum cardinality of a set $S \subseteq X$ such that $\lvert S \cap E_i \rvert \leq 1$ for all i.
$\tau(H)$	transversal number : minimum cardinality of a set $T \subseteq X$ such that $\lvert T \cap E_i \rvert \geq 1$ for all i
$\nu(H)$	matching number : maximum cardinality of a family of pairwise disjoint edges $= \alpha(H^*)$
$q(H)$	chromatic index $= \gamma(H^*)$
$\rho(H)$	covering number $= \tau(H^*)$
$\delta(H)$	maximum degree $= r(H^*)$

This Volume is dedicated

to

PROFESSOR ARNOLD ROSS

ISOMORPHISM PROBLEMS FOR HYPERGRAPHS

Claude Berge, C.N.R.S.

1. Introduction. A hypergraph $H = (X, \mathcal{E}) = (E_1, E_2, \ldots, E_m) = (E_i : i \in M)$ is a family $\mathcal{E}$ of subsets E_i of a set $X = \{x_j : j \in N\}$ of vertices. The sets E_i are called edges.

The rank $r(H)$ of a hypergraph H is the maximum cardinality of the edges. If all edges have the same cardinality, the hypergraph is said to be uniform. The subhypergraph induced by a subset A of X is the hypergraph $H_A = (E_i \cap A : i \in M, E_i \cap A \neq \emptyset)$.

If $I \subseteq M$, the partial hypergraph generated by I is the hypergraph $(E_i : i \in I)$. The section hypergraph is the partial hypergraph $H \times A = (E_i : i \in M, E_i \subseteq A \subseteq X)$.

The dual H^* of H is a hypergraph with vertex set $E = \{e_1, \ldots, e_m\}$, and having edges which are certain subsets of E , namely edges X_j where $X_j = \{e_i : i \in M, x_j \in E_i\}$.

Consider two hypergraphs $H = (E_1, \ldots, E_m)$ and $H' = (F_1, \ldots, F_m)$. H is equivalent to H' $(H \equiv H')$ if the mapping $\varphi : X \to Y$, $\varphi(x_i) = y_i$, satisfies $\varphi(E_i) = F_{\pi i}$ $(i \in M)$ for some permutation π of M .

H is equal to H' (or $H = H'$) if the permutation π in the above definition can be the identity.

H is isomorphic to H' (or $H \cong H'$) if there is a bijection $\varphi : X \to Y$ and if there is a permutation π of M such that $\varphi(E_i) = F_{\pi i}$ $(i \in M)$. The bijection φ is called an isomorphism.

H is strongly isomorphic to H' (or $H \cong H'$) if there is a bijection $\varphi : X \to Y$ for which $\varphi(E_i) = F_i$ for all $i \in M$.

Observe that equality implies the other three relations and any of the relations implies isomorphism.

We give several examples:

Example: Consider the following:

H =

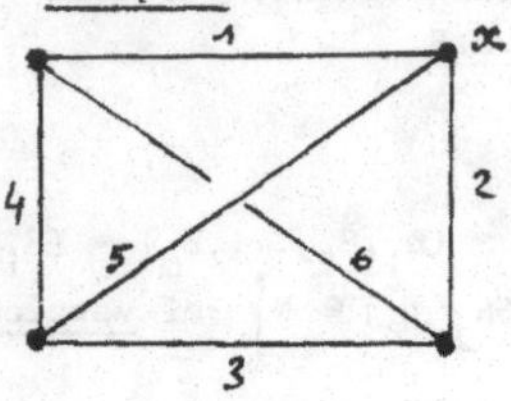

H'=

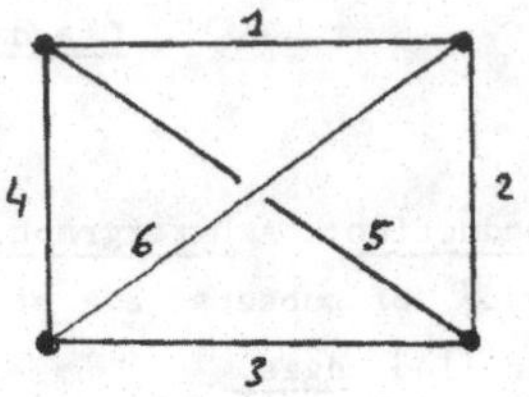

Observe that $H \simeq H'$, but $H \not\cong H'$, since if $H \cong H'$, the vertex x would map to the nonexistent vertex meeting edges 1,2, and 5 in H' .

Example: Consider the line graph L(H) of the graph H above:

L(H) =

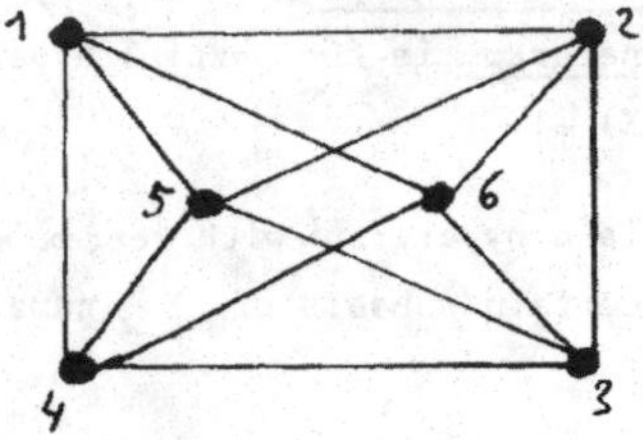

Observe that L(H) = L(H') , but since the edges are unlabeled here, equality is meaningless.

Our purpose in this paper is to present some general results concerning isomorphisms and other relations among hypergraphs.

A *multigraph* is a hypergraph with $|E_i| \leq 2$ for all $i \in M$.

Proposition 1. If $H = (X, (E_i)_{i\in M})$ *and* $H' = (Y, (F_i)_{i\in M})$ *are multigraphs, and if* $\varphi : X \to Y$ *is a bijection, then the following are equivalent:*

(i) φ *is an isomorphism;*

(ii) $m_H(x,y) = m_{H'}(\varphi(x), \varphi(y))$ *for all* $x,y \in X$, *where* $m_H(x,y)$ *is the number of edges joining* x *and* x' *in* H .

Proposition 2. If H *is a hypergraph, then* $(H^*)^* = H$.

Proposition 3. If H, H' *are hypergraphs, then* $H \simeq H'$ *if and only if* $H^* \simeq H'^*$.

Proposition 4. If H, H' *are hypergraphs, then* $H \equiv H'$ *if and only if* $H^* \cong H'^*$.

Proposition 5. The dual of the partial hypergraph of H *generated by* $(E_i : i \in I)$ *equals the subhypergraph of* H^* *induced by* $\{e_i : i \in I\}$.

Proposition 5^* *. The dual of the subhypergraph* $H_A = (E_i \cap A : i \in M, E_i \cap A \neq \emptyset)$ *with* $A = \{x_j : j \in J\}$ *equals the partial hypergraph of* H^* *generated by* $(X_j : j \in J)$.

Proposition 6. The dual of the section hypergraph $H \times A$, $A = \{x_j : j \in J\}$, *equals the subhypergraph of* H^* *induced by* $\bigcup_{j\in N} X_j - \bigcup_{j\in N-J} X_j$.

2. Transitive hypergraphs.

Let $H = (X, \mathcal{E})$ be a hypergraph. Two vertices x and y of H are symmetric if there exists an automorphism φ of H such that $\varphi(x) = y$. Two edges E_i and E_j are symmetric if there exists an automorphism φ of H such that $\varphi(E_i) = E_j$.

H is said to be vertex-transitive (resp. edge-transitive) if any two vertices (resp.edges) are symmetric. A hypergraph that is both vertex-transitive and edge-symmetric is said to be transitive. Because of the duality principle for hypergraphs, the study of vertex-transitive hypergraphs reduces to the study of edge-transitive hypergraphs.

The following result is a generalization of a theorem for graphs due to E. DAUBER [3].

Theorem 1. *For an edge-transitive hypergraph* $H = (X, \mathcal{E})$, *there exists a partition* $(X_1, X_2, \ldots, X_k)$ *of* X *such that*

(1) $\sum_{\lambda} r(X_\lambda) = r(X)$, *where* $r(A)$ *denotes the rank of* H_A,

(2) H_{X_λ} *is transitive for all* λ.

Since H is edge-transitive, $|E_i| = h$ for all i. Let $E_1 = \{x_1, x_2, \ldots, x_h\}$. For $i \in M$, let φ_i be an automorphism such that $\varphi_i(E_1) = E_i$.

Let $Y_p = \{\varphi_i(x_p) \ / \ i \in M\}$ $\quad (p = 1,2,\ldots,h)$
Then, $\overline{H} = (Y_1, \ldots, Y_h)$ is a hypergraph on X, because

$$\bigcup_p Y_p = \bigcup_{i \in M} \varphi_i(E_1) = \bigcup_{i \in M} E_i = X$$

Let $X_1, X_2, \ldots, X_k$ be the connected components of $\overline{H}$.

(1) Let $$E_1^\lambda = \left\{ x_p : p \le h,\ Y_p \subseteq X_\lambda \right\} \quad \text{for } \lambda = 1,2,\ldots,k\ .$$

For $x_p \in E_1^\lambda$,

$$\varphi_i\ (x_p) \in E_i \cap Y_p \subseteq E_i \cap X_\lambda$$

Hence

$$\varphi_i\ (E_1^\lambda) \subseteq E_i \cap X_\lambda$$

Thus,

$$h = \sum_\lambda |E_1^\lambda| = \sum_\lambda |\varphi_i(E_1^\lambda)| \le \sum_\lambda |E_i \cap X_\lambda| = h$$

Hence the equality holds, and

$$E_i \cap X_\lambda = \varphi_i\ (E_1^\lambda)$$

Hence

$$|E_i \cap X_\lambda| = |E_1^\lambda| \qquad (i \in M)$$

This shows that H_{X_λ} is uniform with rank $|E_1^\lambda|$, and furthermore,

$$\sum_\lambda r\ (X_\lambda) = \sum_\lambda |E_1^\lambda| = \sum_\lambda |E_i \cap X_\lambda| = h$$

(2) In H_{X_λ} , the edges $E_i \cap X_\lambda$ and $E_j \cap X_\lambda$ are symmetric, since

$$\varphi_j\ \varphi_i^{-1}\ (E_i \cap X_\lambda) = \varphi_j\ (E_1^\lambda) = E_j \cap X_\lambda\ .$$

Hence H_{X_λ} is edge-transitive.

Furthermore, two vertices $x, y \in Y_p$ are symmetric, since

$$\left.\begin{aligned} x &= \varphi_i\ (x_p) \\ y &= \varphi_j\ (x_p) \end{aligned}\right\} \quad \text{implies} \quad y = \varphi_j\ \varphi_i^{-1}\ (x)$$

Now consider two vertices x, x' in X_λ with $x \in Y_p$, $x' \in Y_q$. There exists a sequence $(Y_p, Y_{p_1}, Y_{p_2}, \ldots, Y_q)$ such that any two consecutive sets of the sequence intersect. Let $x_k \in Y_{p_{k-1}} \cap Y_{p_k}$. In the sequence $(x, x_1, x_2, \ldots, x_q = x')$, any two consecutive vertices are symmetric. Therefore x and x' are symmetric.

Thus, H_{X_λ} is both edge-transitive and vertex-transitive.

Q.E.D.

Corollary 1. If H is an edge-transitive hypergraph that is not vertex-transitive, then H is bicolorable.

This follows, since the partition of X has at least two classes, and they are both transversal sets of H.

Corollary 2. (Dauber) If H is an edge-transitive graph that is not vertex-transitive, then H is bipartite.

This follows from Corollary 1.

3. Extensions of the Whitney Theorem.

Let $G = (E_i : i \in M)$ and $G' = (F_i : i \in M)$ be two connected simple graphs with $|M| = m > 2$. H. Whitney [6] has shown that $|E_i \cap E_j| = |F_i \cap F_j|$ for all i,j implies that $G \cong G'$, unless $G = K_3$ and $G' = K_{1,3}$, or vice versa.

An easy corollary of Whitney's theorem states that if G and G' are two simple graphs different from K_3 and $K_{1,3}$, then $G - E_i \cong G' - F_i$ for all i implies that $G \cong G'$.

(The weak reconstruction conjecture states only that $n(G) > 4$, $G - E_i \cong G' - F_i$ for all i, implies that $G \cong G'$).

In [2], Berge and Rado have proved several extensions of these theorems for hypergraphs.

Denote by $\mathcal{P}(M)$ the set of all subsets of $M = \{1,2,\ldots,m\}$, by $\mathcal{P}_1(M)$ the set of all subsets $I \subseteq M$ such that $|I| \equiv 1$ modulo 2, and by $\mathcal{P}_0(M)$ the set of all subsets $J \subseteq M$ such that $|J| \equiv 0$ modulo 2 and $J \neq \emptyset$. Clearly, $|\mathcal{P}_1(M)| = 2^{m-1}$ and $|\mathcal{P}_0(M)| = 2^{m-1} - 1$ (because the regular bipartite graph whose vertex-sets are $\mathcal{P}_1(M)$ and $\mathcal{P}_0(M) \cup \{\emptyset\}$ and where (S,T) is an edge iff $-1 \leq |S| - |T| \leq 1$, has a perfect matching).

The two <u>Whitney hypergraphs</u> $W_1(M)$ and $W_0(M)$ are defined as follows:
The vertex set of $W_1(M)$ is $\mathcal{P}_1(M)$, and its edges are

$$A_i = \{I : I \in \mathcal{P}_1(M) \, , \, I \ni i\} \qquad (i \in M) \, .$$

The vertex set of $W_0(M)$ is $\mathcal{P}_0(M)$, and its edges are

$$B_i = \{J : J \in \mathcal{P}_0(M) \, , \, J \ni i\} \qquad (i \in M)$$

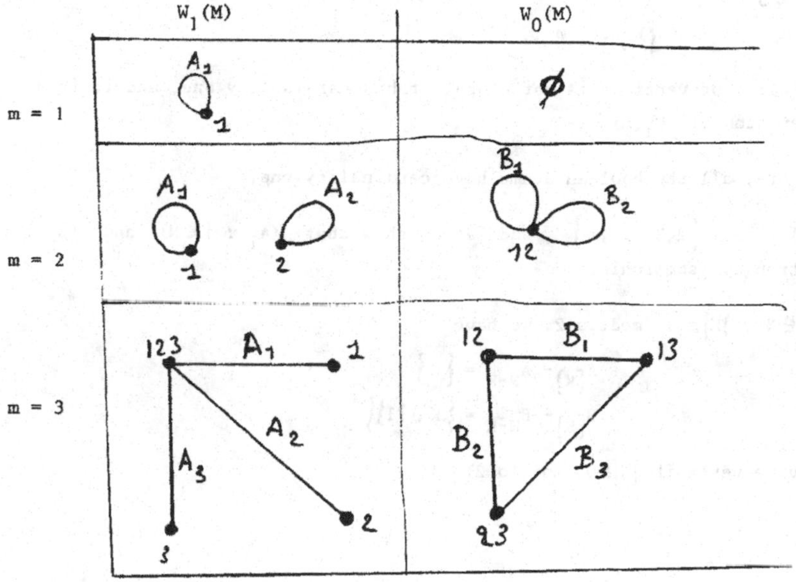

<u>Proposition.</u> <u>For</u> $m \geq 2$, <u>the Whitney hypergraphs</u> $W_1(M) = (A_i : i \in M)$ <u>and</u> $W_0(M) = (B_i : i \in M)$ <u>are two uniform hypergraphs of rank</u> 2^{m-2} ; <u>their boolean atoms have cardinality one, and they are not isomorphic.</u> <u>However, they satisfy</u>

$$W_1(M) - A_i \cong W_0(M) - B_i \qquad (i \in M) .$$

For $K \subseteq M$, $K \neq \emptyset$, put

$$A_K = \bigcup_{i \in K} A_i$$

$$A_{[K]} = \bigcap_{i \in K} A_i$$

$$A_\emptyset = \emptyset$$

Clearly, $W_1(M)$ and $W_0(M)$ are not isomorphic, since $|A_M| = 2^{m-1}$ and $|B_M| = 2^{m-1} - 1$.

For $K \subseteq M$, we have

$$A_{[K]} - A_{M-K} = \{I : I \in \mathcal{P}_1(M) , I \supseteq K\} - \{I : I \in \mathcal{P}_1(M) , I \cap (M-K) \neq \emptyset\}$$
$$= \{K\} \text{ or } \emptyset$$

If this set of vertices is not empty, it has cardinality one, and it is a boolean atom of $W_1(M)$.

Therefore, all the boolean atoms have cardinality one.

Now, let $N = \{2,3,\dots,m\}$, and let us show that $(A_i : i \in N)$ and $(B_i : i \in N)$ are strongly isomorphic.

If $K \subseteq N$, $|K| \equiv 1$ modulo 2, we have

$$A_{[K]} - A_{N-K} = \{K\}$$
$$B_{[K]} - B_{N-K} = \{K \cup \{1\}\}$$

(and vice versa if $|K| \equiv 0$ modulo 2).

Hence, for all $K \subseteq N$, $K \neq \emptyset$, we have

$$\left| A_{[K]} - A_{N-K} \right| = \left| B_{[K]} - B_{N-K} \right| = 1$$

This shows that $(A_i : i \in N) \cong (B_i : i \in N)$.

Q.E.D.

A converse of this proposition is:

<u>Theorem 2.</u> <u>Let</u> $H = (E_i : i \in M)$ <u>and</u> $H' = (F_i : i \in M)$ <u>be two families of subsets</u> $E_i \subseteq X$ and $F_i \subseteq Y$ <u>(with possibly empty edges or infinite edges), with at least one finite edge, such that</u>

(1) <u>for all</u> k , <u>there exists a bijection</u> $\varphi_k : X \to Y$ <u>such that</u> $\varphi_k(E_i) = F_i \quad (i \in M,\ i \neq k)$.

<u>Then</u> $H \cong H'$, <u>unless there exist two sets</u> $A \subseteq X$ <u>and</u> $B \subseteq Y$ <u>with</u> $|A| = |B|$ <u>such that</u> $(A \cap E_i : i \in M) \cong W_\varepsilon(M)$ <u>and</u> $(B \cap F_i : i \in M) \cong W_{1-\varepsilon}(M)$.

The proof, by induction on m , is the same as in [2] , Theorem 2 . For the finite case, a direct proof, shorter than our original one, was found recently by Lovász [5] .

Note that the statement of Theorem 2 would not be true if there is no finite edge: take four infinite sets D_0, D_1, D_2, D_3 of the same cardinality, and put $X = Y = D_0 \cup D_1 \cup D_2 \cup D_3$, $E_1 = D_1 \cup D_2 \cup D_3$, $F_1 = D_2 \cup D_3$, $E_2 = F_2 = D_2$, $E_3 = F_3 = D_3$. $H = (E_1, E_2, E_3)$ and $H' = (F_1, F_2, F_3)$ satisfy (1) , and there is no $A \subseteq X$ such that $(A \cap E_i) \cong W_1(1,2,3)$ and no B such that $(B \cap F_i) \cong W_1(1,2,3)$. However, $H \ncong H'$.

<u>Corollary 1.</u> <u>Let</u> $|M| \geq p \geq 2$, <u>and let</u> $H = (E_i : i \in M)$ <u>and</u> $H' = (F_i : i \in M)$ <u>be two hypergraphs such that</u>

(1) $$(E_i : i \in I) \cong (F_i : i \in I) \qquad (I \subseteq M,\ |I| = p-1)$$

<u>Then</u> $H \cong H'$, <u>unless there exist sets</u> $A \subseteq E_M$, $B \subseteq F_M$ <u>and</u> $P \subseteq M$ <u>such that</u> $|P| = p$ <u>and</u>

$$(2) \quad \begin{cases} (A \cap E_i \ : \ i \in P) \cong W_\varepsilon(P) \\ (B \cap F_i \ : \ i \in P) \cong W_{1-\varepsilon}(P) \end{cases}$$

For $m = p$, consider two hypergraphs H and H' with m edges which satisfy (1) and not (2) .

Let us show first that $|E_M| = |F_M|$; if we have for instance $|E_M| < |F_M|$, consider a set X obtained from E_M by adding $|F_M| - |E_M|$ additional points, and put $Y = F_M$.

By the Theorem 2, there exists a bijection $\varphi : X \to Y$ such that $\varphi(E_i) = F_i$, and therefore

$$\varphi(E_M) = \bigcup F_i = F_M$$

This shows that $|E_M| = |F_M|$ which is a contradiction.

Thus, $|E_M| = |F_M|$, and Theorem 2 , applied with $X = E_M$, $Y = F_M$, shows that $H \cong H'$.

Now, let $m = p+t$, $t \geq 1$, and assume that the statement of this corollary is true for hypergraphs with $p+t-1$ edges. Consider two hypergraphs $H = (E_i : i \in M)$, $H' = (F_i : i \in M)$, with $M = \{1,2,\ldots,m\}$, satisfying (1), and not (2) .

By the induction hypothesis, we have, for $k \in M$,

$$(E_i : i \in M - \{k\}) \cong (F_i : i \in M - \{k\})$$

On the other hand, there exist no sets $A_0 \subseteq E_M$, $B_0 \subseteq F_M$, such that

$$(A_0 \cap E_i : i \in M) \cong W_\varepsilon(M) , \quad (B_0 \cap F_i : i \in M) \cong W_{1-\varepsilon}(M) ,$$

because this would imply the existence of two sets A and B satisfying (2) .

Since the theorem is true for $p = m$, we have also $H \cong H'$.

Q.E.D.

<u>Corollary 2.</u> <u>Let</u> $H = (E_i : i \in M)$ <u>and</u> $H' = (F_i : i \in M)$ <u>be two hypergraphs of rank</u> $h < 2^{p-2}$. <u>If, for every</u> $J \subseteq M$ <u>with</u> $|J| = p-1$, <u>we have</u> $(E_i : i \in J) \cong (F_i : i \in J)$, <u>then</u> $H \cong H'$.

The proof follows immediately from Corollary 1.

<u>Corollary 3.</u> <u>Let</u> $H = (E_i : i \in M)$ <u>and</u> $H' = (F_i : i \in M)$ <u>be two multigraphs such that</u>

(1) $|E_i \cap E_j| = |F_i \cap F_j|$ <u>for all</u> $i,j \in M$

(2) H, H' <u>do not contain as partial graphs</u> $W_\varepsilon(i,j,k)$ <u>and</u> $W_{1-\varepsilon}(i,j,k)$, <u>respectively.</u>

<u>Then</u> $H \simeq H'$.

This follows from Corollary 1 with $p = 3$.

The Whitney Theorem follows easily from Corollary 3 because, if H and H' are connected, of order ≥ 4, and if, say, $(E_i, E_j, E_k) \cong W_1(i,j,k)$ and $(F_i, F_j, F_k) \cong W_0(i,j,k)$, then there exists an edge E_q which has exactly one endpoint in $W_1(i,j,k)$; hence

$$|F_i \cap F_q| + |F_j \cap F_q| + |F_k \cap F_q| = |E_i \cap E_q| + |E_j \cap E_q| + |E_k \cap E_q| = 0 \text{ or } 3 ,$$

which is impossible. If H is of order 4 with more than three edges, it is easy to check that $H \cong H'$.

The following result is in fact due to Lovász [4]

<u>Theorem 3.</u> <u>Let</u> $H = \{E_i : i \in M\}$ <u>and</u> $H' = \{F_i : i \in M\}$ <u>be r-uniform simple hypergraphs of order</u> n <u>with</u> $|M| = m > \frac{1}{2}\binom{n}{r}$, <u>such that</u>

$$H - E_i \cong H' - F_i \qquad (i \in M)$$

<u>Then</u> $H \cong H'$.

Denote by $\bar{H} = \mathcal{P}_r(X) - H$ the complement hypergraph of H, whose number of edges is

$$m(\bar{H}) = \binom{n}{r} - m < m$$

We may assume that $X = Y$. If $\mathcal{S} \subseteq \mathcal{P}_r(X)$, denote by $\alpha(\mathcal{S}, H')$ the number of isomorphisms $\pi : X \to Y$ such that $\{\pi S : S \in \mathcal{S}\} \subseteq \{F_i : i \in M\}$.

By the sieve formula,

$$\alpha(H,H') = \sum_{k=0}^{m} (-1)^k \sum_{\substack{I \subseteq M \\ |I|=k}} \alpha\left(\{E_i : i \in I\}, \bar{H}'\right)$$

Since the terms with $|I| > m(\bar{H})$ are null,

$$\alpha(H,H') = \sum_{k=0}^{m-1} (-1)^k \sum_{\substack{I \subseteq M \\ |I|=k}} \alpha(\{E_i : i \in I\}, \bar{H}') \qquad (1)$$

and

$$\alpha(H',H') = \sum_{k=0}^{m-1} (-1)^k \sum_{\substack{J \subseteq M \\ |J|=h}} \alpha(\{F_j : j \in J\}, \bar{H}') \qquad (2)$$

Since, by hypothesis, H and H' have the same proper partial hypergraphs, the terms in (1) and in (2) are equal, hence:

$$\alpha(H,H') = \alpha(H',H') \geqq 1$$

Q.E.D.

REFERENCES

1. C. Berge, Graphs and Hypergraphs, North Holland Pub. Co., 1973.
2. C. Berge, R.Rado, On isomorphic hypergraphs, and some extensions of Whitney's Theorem to families of sets, J. Comb. Theory B, 13, 1972, 226-241.
3. E. Dauber in F. Harary, Graph Theory, Addison-Wesley, 1969, p. 172.
4. L. Lovász, A Note on the Line Reconstruction Problem, J. Comb. Theory B, 13, 1972, 309-310.
5. L. Lovász, Private communication.
6. H. Whitney, Congruent graphs and the connectivity of graphs, Am. J. Math. 54, 1932, 160-168.

NOMBRES DE COLORATION DE L'HYPERGRAPHE h-PARTI COMPLET

Claude Berge, C.N.R.S.

1. Introduction. Soient n et h deux entiers avec $h \leq n$. L'hypergraphe simple ayant pour ensemble de sommets un ensemble X à n éléments et pour ensemble d'arêtes l'ensemble $\mathcal{P}_h(X)$ des h-parties de X est noté K_n^h et s'appelle l'hypergraphe h-complet à n sommets. Dans le cas $h=2$, on retrouve le graphe complet K_n.

Soit h un entier. L'hypergraphe ayant pour ensemble de sommets la réunion d'ensembles disjoints $X_1, X_2, \ldots, X_h$, avec $|X_i| = n_i$ pour tout i, $1 \leq i \leq h$, et pour arêtes tous les ensembles $\{x_1, x_2, \ldots, x_h\}$ tels que $(x_1, x_2, \ldots, x_h) \in X_1 \times X_2 \times \ldots \times X_h$ est noté $K^h_{n_1, n_2, \ldots, n_h}$ et s'appelle l'hypergraphe h-parti complet sur $\{X_1, X_2, \ldots, X_h\}$. Dans le cas $h=2$, on retrouve le graphe biparti complet K_{n_1, n_2}.

Pour les graphes complets ou les graphes bipartis complets, l'étude de certains coefficients combinatoires comme la classe chromatique est bien connue. L'auteur se propose ici une étude systématique de ces coefficients pour l'hypergraphe h-parti complet.

2. Classe chromatique. Etant donné un hypergraphe $H = (E_i / i \in I)$ on note par :

$\tau(H)$ la cardinalité minimum d'un ensemble transversal de H.

$\tau'(H)$ la cardinalité maximum d'un ensemble transversal minimal de H.

$\nu(H)$ la cardinalité maximum d'un couplage de H.

$q(H)$ l'indice chromatique de H, c'est à dire avec les notations de [1], le nombre chromatique $\gamma(L(H))$ du graphe représentatif $L(H)$.

$s(H)$ le plus grand entier s pour lequel il existe une partition $(I_1, I_2, \ldots, I_s)$ de I telle que les hypergraphes partiels de H $(E_i / i \in I_\alpha)$, pour $1 \leq \alpha \leq s$, soient des recouvrements de l'ensemble des sommets de H. Ce nombre a été considéré par Bernt Lindström dans [2].

$\delta(H)$ le degré maximum de H, c'est à dire le nombre maximum d'arêtes de H contenant un même sommet de H.

$\delta'(H)$ <u>le degré minimum</u> de H, c'est à dire le nombre minimum d'arêtes de H contenant un même sommet de H.

On appelle <u>clique</u> de H un hypergraphe partiel $H' = (E_j/j \in J)$ de H tel que $E_k \cap E_\ell \neq \emptyset$ pour k et ℓ dans J.

On note par :

$t(H)$ <u>le plus petit entier t pour lequel il existe une partition $(I_1,I_2,\dots,I_t)$ de I telle que les hypergraphes partiels de H $(E_i/i \in I_\alpha)$, pour $1 \leq \alpha \leq t$, soient des cliques de</u> H.

Avec les notations de [1] on a $t(H) = \theta(L(H))$.

$\omega(H)$ <u>le plus grand nombre d'arêtes d'une clique de</u> H.

Rappelons que si $H = K_n^h$, P. Erdös, Chao Ko et R. Rado [3] ont montré que

$$\omega(K_n^h) = \begin{cases} \binom{n}{h} & \text{si } n < 2h \\ \binom{n-1}{h-1} & \text{si } n \geq 2h. \end{cases}$$

Par contre, $q(K_n^h)$ n'est pas connu excepté le cas h=2 du à E. Lucas [4] et le cas h=3 du à R. Peltesohn [5] et $t(K_n^h)$ n'est connu que dans les cas h=2 et h=3 dus à J.C. Meyer [6].

Considérons sur $\{X_1,X_2,\dots,X_h\}$ l'hypergraphe h-parti complet $K^h_{n_1,n_2,\dots n_h}$ avec $n_1 \leq n_2 \leq \dots \leq n_h$. Pour cet hypergraphe, on a les résultats suivants.

<u>Théorème 1. Les seuls ensembles transversaux minimaux de</u> $K^h_{n_1,n_2,\dots,n_h}$ <u>sont les ensembles</u> $X_1,X_2,\dots,X_h$ <u>; en particulier, on a</u> :

$$\tau(K^h_{n_1,n_2,\dots n_h}) = n_1$$

$$\tau'(K^h_{n_1,n_2,\dots,n_h}) = n_h.$$

En effet, soit $T \subset X = \bigcup_{i=1}^{h} X_i$ un ensemble transversal des arêtes de l'hypergraphe et supposons qu'il ne contienne aucun des ensembles X_i. Pour $i=1,2,\dots,h$, considérons un sommet $a_i \in X_i - T$. Alors $\{a_1,a_2,\dots,a_h\}$ est une arête de l'hypergraphe qui ne rencontre pas T, ce qui contredit la définition d'un ensemble transversal.

Comme d'autre part les ensembles X_i sont évidemment des ensembles transversaux de l'hypergraphe, l'énoncé s'ensuit.

Théorème 2. On a :

$$\nu(K^h_{n_1,n_2,\dots,n_h}) = n_1.$$

En effet si $X_j = \{x^i_j \mid 1 \le i \le n_j\}$ pour $1 \le j \le h$, les arêtes $\{x^i_1, x^i_2, \dots, x^i_h\}$ pour $1 \le i \le n_1$, constituent un couplage ayant n_1 arêtes. Par suite, $\nu(K^h_{n_1,n_2,\dots,n_h}) \ge n_1$. D'autre part, puisque pour tout hypergraphe H on a $\nu(H) \le \tau(H)$, on obtient, d'après le théorème 1, $\nu(K^h_{n_1,n_2,\dots n_h}) \le n_1$

C.Q.F.D.

Théorème 3. L'indice chromatique $q(K^h_{n_1,n_2,\dots,n_h})$ est égal au degré maximum de $K^h_{n_1,n_2,\dots,n_h}$, c'est à dire à :

$$\delta(K^h_{n_1,n_2,\dots n_h}) = n_2 n_3 \dots n_h.$$

Si ζ désigne un p-uple $(x_1, x_2, \dots, x_p) \in X_1 \times X_2 \times \dots \times X_p$ avec $p \le h$, on notera $\tilde{\zeta}$ l'ensemble $\{x_1, x_2, \dots, x_p\}$.

Il est évident que le degré maximum de $K^h_{n_1,n_2,\dots,n_h}$ est $n_2 n_3 \dots n_h$. Puisque $q(H) \ge \delta(H)$ pour tout hypergraphe H, il suffit de démontrer qu'il existe une coloration des arêtes de $K^h_{n_1,n_2,\dots,n_h}$ en $n_2 n_3 \dots n_h$ couleurs telle que deux arêtes de la même couleur soient disjointes.

Pour h=2, le théorème de D. König ([1],p. 238) implique que $q(K_{n_1,n_2}) = \delta(K_{n_1,n_2})$ et l'énoncé est vérifié. Soit $h > 2$. Supposons l'énoncé vérifié pour tous les hypergraphes h'-parti complets avec $h' < h$, et démontrons-le pour $K^h_{n_1,n_2,\dots,n_h}$. En vertu de l'hypothèse d'induction, l'hypergraphe $K^{h-1}_{n_1,n_2,\dots,n_{h-1}}$ admet une coloration en $n_2 n_3 \dots n_{h-1}$ couleurs. Donc il existe une application φ de $X_1 \times X_2 \times \dots \times X_{h-1}$ sur $[0,n_2-1] \times [0,n_3-1] \times \dots \times [0,n_h-1]$, où $[0,n_{h-1}-1]$ désigne les entiers ≥ 0 et $\le n_j - 1$, telle que :

$$\left.\begin{array}{l} \zeta, \zeta' \in X_1 \times X_2 \times \dots \times X_h \\ \zeta \ne \zeta' \ , \ \tilde{\zeta} \cap \tilde{\zeta}' \ne \emptyset \end{array}\right\} \text{ implique } \varphi(\zeta) \ne \varphi(\zeta').$$

Pour tout $\gamma \in [0,n_2-1] \times [0,n_3-1] \times \ldots \times [0,n_{h-1}-1]$, posons :

$$C_\gamma = \{\zeta \in X_1 \times X_2 \times \ldots \times X_{h-1} \mid \varphi(\zeta) = \gamma\}.$$

Autrement dit, C_γ est l'ensemble des arêtes de $K^{h-1}_{n_1,n_2,\ldots,n_{h-1}}$ ayant la couleur γ. Comme les arêtes de couleur γ constituent un couplage, on a, d'après le théorème 2, $|C_\gamma| \leq n_1$. Par conséquent on peut, pour tout $\gamma \in [0,n_2-1] \times [0,n_3-1] \times \ldots \times [0,n_{h-1}-1]$ et pour tout $\zeta \in C_\gamma$, associer un entier $f(\zeta)$, $0 \leq f(\zeta) \leq n_1-1$, de sorte que la restriction de f à C_γ soit une <u>injection</u> de C_γ dans $[0,n_1-1]$. De même soit g une bijection de X_h sur $[0,n_h-1]$.

Posons, pour $(x_1,x_2,\ldots,x_h) \in X_1 \times X_2 \times \ldots \times X_h$,

$$\varphi_o(x_1,x_2,\ldots,x_{h-1},x_h) = (\varphi(x_1,x_2,\ldots,x_{h-1}),[f(x_1,x_2,\ldots,x_{h-1})+g(x_h)]_{n_h})$$

où $[\alpha]_\lambda$ désigne l'entier entre 0 et $\lambda-1$ congru à α modulo λ. φ_o est une fonction définie sur l'ensemble des arêtes de $K^h_{n_1,n_2,\ldots,n_h}$ et qui peut prendre au plus $n_2 \ldots n_{h-1} n_h$ valeurs différentes. Il s'agit maintenant de montrer que c'est une coloration des arêtes de $K^h_{n_1,n_2,\ldots n_h}$, c'est à dire que :

$$\left.\begin{array}{l}(x_1,x_2,\ldots,x_h) \in X_1 \times X_2 \times \ldots \times X_h \\ (x'_1,x'_2,\ldots,x'_h) \in X_1 \times X_2 \times \ldots \times X_h \\ \{x_1,x_2,\ldots,x_h\} \cap \{x'_1,x'_2,\ldots,x'_h\} \neq \emptyset\end{array}\right\} \text{implique } \varphi_o(x_1,x_2,\ldots,x_h) \neq \varphi_o(x'_1,x'_2,\ldots,x'_h)$$

Posons $\zeta = (x_1,x_2,\ldots,x_{h-1})$ et $\zeta' = (x'_1,x'_2,\ldots,x'_{h-1})$ et considérons les trois cas suivants :

<u>Cas 1</u>.

$$\zeta = \zeta' \text{ et } x_h \neq x'_h.$$

Dans ce cas, $[f(\zeta) + g(x_h)]_{n_h} \neq [f(\zeta) + g(x'_h)]_{n_h}$ et par suite $\varphi_o(\zeta,x_h) \neq \varphi_o(\zeta',x'_h)$.

<u>Cas 2</u>.

$$\zeta \neq \zeta' \text{ et } x_h = x'_h.$$

Si $\varphi(\zeta) \neq \varphi(\zeta')$ on a $\varphi_o(\zeta,x_h) \neq \varphi_o(\zeta',x'_h)$.

Si au contraire, $\varphi(\zeta) = \varphi(\zeta')$, on a $f(\zeta) \neq f(\zeta')$; puisque $0 \leq f(\zeta) \leq n_h-1$, $0 \leq f(\zeta') \leq n_h-1$, on a aussi

$$\varphi_o(\zeta,x_h) \neq \varphi_o(\zeta',x'_h).$$

Cas 3.

$\zeta \neq \zeta'$ et $x_h \neq x'_h$.

Dans ce cas $\tilde{\zeta} \cap \tilde{\zeta} \neq \emptyset$, donc $\varphi(\zeta) \neq \varphi(\zeta')$ et par suite

$$\varphi_o(\zeta, x_h) \neq \varphi_o(\zeta', x'_h).$$

C.Q.F.D.

Corollaire 1. On a : $\omega(K^h_{n_1,n_2,\ldots,n_h}) = n_2 n_3 \ldots n_h$.

En effet, on a, pour tout hypergraphe H, $q(H) \geq \omega(H)$. Il s'agit donc de montrer que $\omega(K^h_{n_1,n_2,\ldots,n_h}) \geq q(K^h_{n_1,n_2,\ldots,n_h})$, c'est à dire qu'il existe une clique ayant $q(K^h_{n_1,n_2,\ldots,n_h}) = n_2 n_3 \ldots n_h$ arêtes. Une telle clique est obtenue en prenant toutes les arêtes incidentes à un sommet $x_1 \in X_1$.

Corollaire 2. On a : $t(K^h_{n_1,n_2,\ldots n_h}) = n_1$.

Remarquons d'abord que, pour tout hypergraphe $H = (E_i / i \in I)$, on a $t(H) \leq \tau(H)$, car si $T = \{a_1, a_2, \ldots, a_k\}$ est un transversal minimum de H on peut prendre pour I_1 tous les indices $i \in I$ tels que $a_1 \in E_i$, pour I_2 tous les indices $i \in I$ tels que $a_2 \in E_i$ et $a_1 \notin E_i$ et de façon générale, pour I_j, $1 \leq j \leq k$, tous les indices $i \in I$ tels que $a_j \in E_i$ et $a_\ell \notin E_i$ pour $\ell < j$. On obtient ainsi une partition $(I_1, I_2, \ldots, I_k)$ de I en $k = \tau(H)$ classes.

Donc d'après le théorème 1 :

$$t(K^h_{n_1,n_2,\ldots,n_h}) \leq \tau(K^h_{n_1,n_2,\ldots,n_h}) = n_1$$

D'autre part, si G est un graphe d'ordre n, on a pour une partition minimum $(C_1, C_2, \ldots, C_k)$ en $k = \theta(G)$ cliques :

$$n = |C_1| + |C_2| + \ldots + |C_k| \leq \theta(G)\ \omega(G)$$

Donc :

$$t(K^h_{n_1,n_2,\ldots,n_h}) = \theta(L(K^h_{n_1,n_2,\ldots,n_h})) \geq \frac{n_1 n_2 \ldots n_h}{\omega(K^h_{n_1,n_2,\ldots,n_h})}$$

et par suite, d'après le corollaire 1,

$$t(K^h_{n_1,n_2,\dots,n_h}) \geq \frac{n_1 n_2 \dots n_h}{n_2 \dots n_h} = n_1.$$

L'égalité s'ensuit. C.Q.F.D.

<u>Théorème 4.</u> <u>On a</u> : $s(K^h_{n_1,n_2,\dots n_h}) = n_1 n_2 \dots n_{h-1}$.

En effet, pour tout hypergraphe $H = (E_i / i \in I)$, on a $s(H) \leq \delta'(H)$, car si $(I_1, I_2, \dots, I_s)$, $s = s(H)$, est une partition de I telle que les hypergraphes partiels $(E_i / i \in I_\alpha)$, avec $1 \leq \alpha \leq s$ soient des recouvrements de H, pour tout sommet x de H et pour tout α, $1 \leq \alpha \leq s$, il existe un indice $i \in I_\alpha$ tel que $x \in E_i$. On a donc

$$s(K^h_{n_1,n_2,\dots n_h}) \leq \delta'(K^h_{n_1,n_2,\dots,n_h}) = n_1 n_2 \dots n_{h-1}.$$

Nous allons montrer maintenant qu'il existe $n_1 n_2 \dots n_{h-1}$ hypergraphes partiels de $K^h_{n_1,n_2,\dots n_h}$ ayant chacun n_h arêtes recouvrant les sommets de $K^h_{n_1,n_2,\dots,n_h}$, deux quelconques d'entre eux n'ayant pas d'arête commune.

Si $h=2$, ceci découle d'un théorème de R.P. Gupta [7]. On peut le voir directement de la façon suivante. Soit $Y \subset X_2$ avec $|Y| = n_1$. Il existe d'après le théorème de D. König ([1] p. 238) n_1 couplages parfaits $C_1, C_2, \dots, C_{n_1}$ partitionnant les arêtes du graphe biparti complet K_{n_1,n_1}. Chaque point x de $X_2 - Y$ (s'il y en a) étant de degré n_1, il existe une bijection f_x de l'intervalle $[1, n_1]$ des entiers ≥ 1 et $\leq n_1$ sur l'ensemble des arêtes de K_{n_1,n_2} incidentes à x. On obtient alors les n_1 graphes partiels répondant à la question en prenant

$$H_i = C_i \cup \{f_x(i) | \ x \in X_2 - Y\} \quad \text{pour} \quad 1 \leq i \leq n_1.$$

Le nombre d'arêtes de H_i est égal à :

$$|C_i| + |\{f_x(i)| \ x \in X_2 - Y\}| = n_1 + n_2 - n_1 = n_2.$$

Soit $h > 2$, et supposons le résultat établi pour l'hypergraphe $(h-1)$-parti $K^{h-1}_{n_1,n_2,\dots,n_{h-1}}$.

Il existe donc $n_1 n_2 \dots n_{h-2}$ hypergraphes partiels H'_i, $1 \leq i \leq n_1 n_2 \dots n_{h-2}$ de $K^{h-1}_{n_1,n_2,\dots,n_{h-1}}$ chacun ayant n_{h-1} arêtes, recouvrant les sommets de

$K^{h-1}_{n_1,n_2,\dots n_{h-1}}$, et deux à deux disjoints.

Soient $\mathcal{E}^i = \{E^i_j / 1 \le j \le n_{h-1}\}$ les ensembles d'arêtes de ces recouvrements disjoints, et posons :

$$X'_h = \{1,2,\dots n_{h-1}\}\ ,\ X''_h = \{n_{h-2},\dots,n_h\},$$

$$X_h = X'_h \cup X''_h\ .$$

Considérons les n_{h-1} permutations circulaires φ_k de X'_h définies par $\varphi_k(1) = k$.

Chaque point x de X''_h étant de degré $n_1 n_2 \dots n_{h-1}$ il existe une bijection f_x de $[1, n_1 n_2 \dots n_{h-2}] \times [1, n_{h-1}]$ sur l'ensemble des arêtes de $K^h_{n_1,n_2,\dots n_h}$ incidentes à x. On obtiendra les $n_1 n_2 \dots n_{h-1}$ hypergraphes partiels $H_{i,k}$ répondant à la question en posant, pour $1 \le i \le n_1 n_2 \dots n_{h-2}$ et $1 \le k \le n_{h-1}$,

$$H_{i,k} = \{E^i_j \cup \{\varphi_k(j)\} / 1 \le j \le n_{h-1}\} \cup \{f_x(i,k) \ / \ x \in X''_h\}$$

C.Q.F.D.

Avant d'énoncer le théorème suivant, rappelons qu'un hypergraphe H est dit <u>héréditaire</u> si tout sous-ensemble non vide d'une arête de H est une arête de H ; on dénote par $\hat{H}$ la fermeture héréditaire d'un hypergraphe non héréditaire.

<u>Théorème 5</u>. <u>Soit</u> $H = \hat{K}^h_{n_1,n_2,\dots,n_h}$ <u>l'hypergraphe héréditaire dont les arêtes maximales sont celles de l'hypergraphe</u> $K^h_{n_1,n_2,\dots,n_h}$. <u>L'indice chromatique</u> $q(H)$ <u>est égal au degré maximum de</u> H, <u>c'est à dire à</u>

$$\delta(H) = (n_2+1)\ (n_3+1)\ \dots\ (n_h+1).$$

Soit $\{a_1, a_2, \dots a_h\}$ un ensemble auxiliaire de cardinalité h disjoint de l'ensemble des sommets de H.

A toute arête E de H associons l'ensemble

$$E' = E \cup \{a_i \mid 1 \le i \le h\ ,\ E \cap X_i = \emptyset\}.$$

On réalise ainsi une bijection entre l'ensemble des arêtes de H et l'ensemble des arêtes de $K^h_{n_1+1\,,\,n_2+1,\ldots,n_h+1}$ différentes de $\{a_1,a_2,\ldots a_h\}$

Le degré d'un point $x \in X_1$, est le même dans H et dans l'hypergraphe $K^h_{n_1+1\,,\,n_2+1\,,\ldots,\,n_h+1}$, et, par suite,

$$\delta(H) = (n_2+1)\ (n_3+1)\ \ldots\ (n_h+1).$$

Donc

$$q(H) \geq (n_2+1)(n_3+1)\ \ldots\ (n_h+1).$$

D'autre part, H étant un sous-hypergraphe de $K^h_{n_1+1,n_2+1,\ldots,n_h+1}$, on a

$$q(H) \leq q(K^h_{n_1+1,n_2+1,\ldots,n_h+1})$$

D'après le théorème 3, on a donc

$$q(H) \leq (n_2+1)(n_3+1)\ldots(n_h+1)$$

Le résultat est ainsi prouvé. C.Q.F.D.

REFERENCES

1. C. Berge, Graphes et Hypergraphes, Dunod, Paris, 1970

2. B. Lindström, A Theorem on families of Sets, J. of Combinatorial Theory, A. 13, 1972, 274-277.

3. P. Erdös, Chao Ko, R. Rado, Intersection theorems for systems of finite sets, Quart. J. of Math. (Oxford) (2) 12 (1961), 313-320.

4. E. Lucas, Récréations Mathématiques, A. Blanchard, Paris, 1892-1924.

5. R. Peltesohn, Das Turnier Problem für Spiele zu je dreier, Inaugural Dissertation, Friedrich-Wilhelms-Universität zu Berlin 1936.

6. J.C. Meyer, Quelques problèmes concernant les cliques des hypergraphes h-complets et q-parti h-complets. Working Seminar on Hypergraphs, Columbus Ohio (1972), Springer-Verlag.

7. R.P. Gupta, A decomposition theorem for bipartite graphs, Théorie des graphes, Rome I.C.C. (P. Rosenstiel ed.), Dunod, Paris, 1967, 135-138.

THE COLORING NUMBERS OF THE DIRECT PRODUCT OF TWO HYPERGRAPHS

C. Berge, University of Paris VI

M. Simonovits, Eötvös L. University, Budapest

1. Definitions. In the following, $H = (X, \mathcal{E})$ will denote a hypergraph with vertex set $X = \{x_1, x_2, \ldots, x_n\}$, and edge family $\mathcal{E} = (E_i \,/\, i \in I)$. $n(H) = n$ is the order of H, $m(H) = |I|$ is the number of edges, and $r(H) = \max|E_i|$ is the rank of H. A set $S \subseteq X$ is said to be stable if it contains no edge; the maximum cardinal of a stable set is denoted by $\beta(H)$ and is called the stability number of H.

A set $T \subset X$ is said to be a transversal if it meets each edge; the minimum cardinal of a transversal of H is denoted by $\tau(H)$ and is called the transversal number of H. Other numbers can be associated with hypergraph H; for instance, $\nu(H)$ denotes the maximum number of pairwise disjoint edges; $\rho(H)$ denotes the minimum number of edges which together cover X; $\delta(H)$, the maximum degree, is the maximum number of edges which meet at the same vertex. $\chi(H)$, the chromatic number, is the least integer k for which there exists a partition of X into k stable sets.

It is well known that the following inequalities hold:

$$(1)\quad \chi(H)\,\beta(H) \geq n(H)$$

$$(2)\quad \chi(H) + \beta(H) \leq n(H) + 1$$

$$(3)\quad \beta(H) = n(H) - \tau(H)$$

$$(4)\quad \tau(H) \geq \nu(H)$$

$$(5)\quad \tau(H) \leq r(H)\,\nu(H)$$

(For a proof see [1]).

Given two hypergraphs $H = (X, \mathcal{E})$ and $H' = (Y, \mathcal{F})$, with $\mathcal{E} = (E_i \,/\, i \in I)$, $\mathcal{F} = (F_j \,/\, j \in J)$, their direct product is a hypergraph $H \times H'$ with vertex set $X \times Y$ and with edges $E_i \times F_j$ for $(i, j) \in I \times J$.

The aim of this paper is to find upper bounds and lower bounds for the numbers associated with hypergraph $H \times H'$. These results can easily be extended to the direct product of more than two hypergraphs.

First, it should be noticed that we have:

$$(6) \quad r(H \times H') = r(H)\, r(H')$$

Moreover, some of the associated numbers of $H \times H'$ can be obtained from other coefficients by the duality principle, using the following result:

Proposition 1. $(H \times H')^* = H^* \times H'^*$

By definition of the dual, $(H \times H')^*$ has vertex set $\{(e_i, f_j) / i \in I, j \in J\}$; the edge corresponding to a vertex (x_p, y_q) of $H \times H'$ must contain all the (e_i, f_j) such that $E_i \ni x_p$ and $F_j \ni y_q$, and therefore is the set $X_p \times Y_q$. Here X_p is the set of e_i's such that $E_i \ni x_p$, x_q is similarly defined. Hence the edge family of $H \times H'$ is

$$\{(X_p, X_q) / 1 \leq p \leq m, 1 \leq q \leq h\}.$$

The proposition follows.

2. The Transversal Number. Let H and H' be two hypergraphs of order m and n, respectively. From (3) we have

$$\beta(H \times H') = mn - \tau(H \times H').$$

So, the problem of finding a lower bound for β is the same as the problem of finding an upper bound for τ. This problem often occurs in Combinatorics.

Example 1. What is the least number of points in a $m \times n$ rectangular unit lattice (integer points of the plane), such that each square of side r has at least one of these points as a corner? The answer is $\tau(D_m^r \times D_n^r)$, where D_n^r is a simple graph with vertices $1, 2, \ldots, n$, two vertices x, y being joined if $|x - y| = r$.

One can easily show that if $r = 1$ and mn is even, we have

$$\tau(D_m^1 \times D_n^1) = [m/2]^* \, [n/2]^*,$$

where $[x]^*$ denotes the smallest integer $\geq x$.

Example 2. The Zarankiewicz problem. Let $1 \leq r \leq m$, $1 \leq s \leq n$. Zarankiewicz has asked for the least integer $k_{rs}(m, n)$ such that every subset of $k_{rs}(m, n)$ points of an $m \times n$ rectangular unit lattice should contain rs points situated in r columns and s rows. If K_m^r denotes the complete r - uniform hypergraph on m points, we have

$$\beta(K_m^r \times K_n^s) = k_{rs}(m, n) - 1 \quad .$$

An extensive literature exists on this problem (see Guy, [7], [8]). For the sake of simplicity, consider first the case $m = n$. It is known [9] that if $r \leq s$, then

$$\text{(i)} \qquad \beta(K_n^r \times K_n^s) \leq c_{r,s}\, n^{2 - 1/r}$$

where $c_{r,s}$ is a constant. Furthermore, if $r = s = 2$, (i) is sharp, that is if $n \to \infty$, we have

$$\frac{\beta(K_n^2 \times K_n^2)}{n^{3/2}} \to 1$$

It follows easily from [2] that if $s \geq 3$, then

$$\text{(ii)} \qquad c'_s\, n^{5/3} \leq \beta(K_n^3 \times K_n^s) \leq c''_s\, n^{5/3}$$

Unfortunately, the lower bounds for the general case are far from the upper bound given in (i).

Another simple case is when n is much greater than m . Thus if $n \geq (s - 1)\binom{m}{r}$, Čulik [5] has determined the exact value:

$$\beta(K^r \times K_n^s) = (r - 1)n + (s - 1)\binom{m}{r} \quad .$$

For example, $\beta(K_4^2 \times K_6^2) = 6 + 6 = 12$, and a maximum stable set with 12 vertices is given by the ones in the following array:

$$n = 4 \left\{ \underbrace{\begin{matrix} 1 & 1 & 1 & 0 & 0 & 0 \\ 0 & 0 & 1 & 1 & 1 & 0 \\ 1 & 0 & 0 & 0 & 1 & 1 \\ 0 & 1 & 0 & 1 & 0 & 1 \end{matrix}}_{m = 6} \right.$$

<u>Proposition 2</u>. <u>Let</u> H <u>and</u> H' <u>be two hypergraphs</u>. <u>Then</u>

$$\tau(H \times H') \leq \tau(H)\,\tau(H') \quad .$$

Let $T \subset X$ and $T' \subset Y$ be two minimum transversals respectively for H and H'. Since $T \times T'$ is a transversal for $H \times H'$, we have

$$\tau(H \times H') \leq |T \times T'| = \tau(H)\tau(H') \; .$$

Q.E.D.

Instead of showing that the inequality of Proposition 1 is the best possible, we shall show that a very large class of hypergraphs H satisfy

$$\tau(H \times H') = \tau(H)\,\tau(H') \qquad \text{for all } H' \; .$$

First, we shall prove two lemmas. In fact, these lemmas have been proved independently by L. Lovász and the authors and can be used for a different purpose (see [10]). Let s be a positive integer. Let $\varphi(x)$ be an integer function on X ; for $A \subset X$, let

$$\varphi(A) = \sum_{x \in A} \varphi(x) \; .$$

If $\varphi(E_i) \geq s$ for all $i \in I$, the function φ is said to be an <u>s-covering</u> for H. The minimum of $\varphi(X)$ over all s-coverings φ will be denoted by $\tau_s(H)$. Clearly, $\tau(H) = \tau_1(H)$.

Now, let H be a hypergraph with vertices $x_1, x_2, \ldots, x_n$, with m(H) edges, and with maximum degree $\delta(H)$. Let $\alpha_1, \alpha_2, \ldots, \alpha_n$ be n non-negative real numbers. Let

$$\tau^*(H) = \min\Big\{ \sum_{i=1}^{n} \alpha_i \;/\; \sum_{x_i \in E_j} \alpha_i \geq 1 \quad \text{for all } j \Big\}$$

<u>Lemma 1</u>. <u>Let</u> H <u>be a hypergraph</u>. <u>Then</u>

$$\max\Big\{\nu(H)\;,\;\frac{m(H)}{\delta(H)}\Big\} \leq \tau^*(H) \leq \frac{\tau_s(H)}{s} \leq \tau(H)$$

We have $\tau_s(H) \leq s\,\tau(H)$, because if T is a minimum transversal set and if $\varphi_T(X)$ is its characteristic function, then $s\varphi_T$ is an s-covering, and, consequently,

$$\tau_s(H) \leq s\varphi_T(X) = s\,\tau(H) \quad .$$

We have $\frac{1}{s}\tau_s(H) \geq \tau^*(H)$, because if φ is a minimum s-covering, then by putting $\alpha_i = \frac{1}{s}\varphi(x_i)$, we obtain

$$\tau^*(H) \leq \sum_{i=1}^{n} \alpha_i = \frac{1}{s}\tau_s(H) \quad .$$

Now, we shall show that $\tau^*(H) \geq \frac{m(H)}{\delta(H)}$. Consider n real numbers α_i such that $\sum_{x_i \in E_j} \alpha_i \geq 1$ for all j and such that $\Sigma\alpha_i = \tau^*(H)$. Denote by $\delta_x(H)$ the degree of vertex x . We have

$$m(H) \leq \sum_{j=1}^{m} \sum_{x_i \in E_j} \alpha_i \leq \sum_{i=1}^{n} \alpha_i \delta_{x_i}(H) \leq \delta(H) \sum_{i=1}^{n} \alpha_i = \delta(H)\,\tau^*(H)$$

Also, if $(E'_1, E'_2, \ldots, E'_\nu)$ is a maximum matching of H , then

$$\tau^*(H) = \sum^{n} \alpha_i \geq \sum_{k=1}^{\nu} \sum_{x_i \in E'_k} \alpha_i \geq \nu(H)$$

The first inequality follows.

Lemma 2. $s^{-1}\tau_s(H)$ tends to a limit, and

$$\lim_{s \to \infty} \frac{\tau_s(H)}{s} = \tau^*(H)$$

A well known theorem of Fekete states that if a sequence (a_n) of positive numbers is such that $a_{m+n} \leq a_m + a_n$, then the sequence $(\frac{a_n}{n})$ tends to a limit. Let φ be a minimum p-covering and φ' be a minimum q-covering. Then $\varphi + \varphi'$ is a $(p+q)$-covering, and therefore

$$\tau_{p+q}(H) \leq \varphi(X) + \varphi'(X) = \tau_p(H) + \tau_q(H) \quad .$$

Hence, by Fekete's theorem, there exists a number ξ such that $\frac{\tau_s(H)}{s} \to \xi$. By Lemma 1, $\xi \geq \tau^*(H)$.

Furthermore, the α_i's whose sum is $\tau^*(H)$ are defined by a linear programming problem with integral coefficients, and therefore, the α_i's are rational, and we can write

$$\alpha_i = \frac{\alpha'_i}{s} , \; \alpha'_i \text{ and } s \text{ integers.}$$

Hence

$$\frac{\tau_s(H)}{s} \leq \frac{1}{s} \Sigma\alpha'_i = \Sigma\alpha_i = \tau^*(H)$$

This shows that $\xi = \tau^*(H)$.

Q.E.D.

Theorem 1. A necessary and sufficient condition for a hypergraph H to satisfy $\tau(H \times H') = \tau(H)\tau(H')$ for all H' is that $\tau(H) = \tau^*(H)$.

Necessity. Assume that $\tau(H) \neq \tau^*(H)$. Then, by Lemma 1, $\tau(H) > \tau^*(H)$ and by Lemma 2, there exists an integer $s > 2$ such that $\frac{\tau_s(H)}{s} < \tau(H)$. We shall show that there exists a hypergraph H' such that $\tau(H \times H') < \tau(H)\tau(H')$.

Let $\varphi(x)$ be a minimal s-covering for H . Put $\varphi(X) = \tau_s(H) = t$, $Y = \{1, 2, \ldots, t\}$.

It is always possible to associate with each $x \in X$ a set $A(x) \subset Y$ so that:

(1) $|A(x)| = \varphi(x)$ for all $x \in X$,

(2) $x \neq x'$ implies $A(x) \cap A(x') = \emptyset$.

Let $H' = K_t^{t-s+1} = (Y, (F_j))$. We shall show that the direct product $H \times H'$ admits

$$T_o = \{(x, y) / x \in X, y \in A(x)\}$$

as a transversal.

Clearly, $E_i \times Y$ contains at least s different elements of T_o . Since no two of them have the same projection on Y , $E_i \times F_j$ contains at least one element of T_o , for all i, j . Thus, T_o is a transversal of $H \times H'$. Moreover, $\tau(H') = s$. Hence

$$\tau(H \times H') \leq |T_o| = \tau_s(H) < s\tau(H) = \tau(H)\tau(H') .$$

Q.E.D.

Sufficiency. Let H be a hypergraph such that $\tau(H) = \tau^*(H)$. Then by Lemma 1, $\tau_s(H) = s\tau(H)$ for every integer s . Let $T_o \subset X \times Y$ be a minimum transversal of $H \times H'$. Let

$$\varphi_o(x) = |\{y / (x, y) \in T_o, y \in Y\}| .$$

Since the projection on Y of $(E_i \times Y) \cap T_o$ is a transversal of H' ,

$$\varphi_o(E_i) = |(E_i \times Y) \cap T_o| \geq \tau(H') .$$

Thus, ϖ_o is an s-covering for $s = \tau(H')$. Hence,

$$\tau(H \times H') = |T_o| = \varpi_o(X) \geq \tau_s(H) = s\,\tau(H) = \tau(H')\,\tau(H) \ .$$

Therefore, the equality holds.

Q.E.D.

Corollary 1. If H satisfies $\nu(H) = \tau(H)$ (and in particular if H is balanced) then $\tau(H \times H') = \tau(H)\,\tau(H')$ for every H' .

This follows immediately from Lemma 1.

In particular, if H is balanced, i.e. if each odd cycle of H possesses an edge containing three vertices of the cycle, it is known ([1]) that $\nu(H) = \tau(H)$, and consequently, the required equality holds.

Corollary 2. Let G be a graph. Then

$$\tau(G \times H') = \tau(G)\,\tau(H')$$

for every hypergraph H' if and only if

$$\tau(G) = \nu(G) \ .$$

By a theorem of Lovász [10], $\tau(G) = \tau^*(G)$ if and only if $\tau(G) = \nu(G)$. The proof follows.

Q. E. D.

Corollary 3. Let H be a hypergraph such that $m(H) = \tau(H)\,\delta(H)$. Then $\tau(H \times H') = \tau(H)\,\tau(H')$ for every hypergraph H' .

This follows immediately from Lemma 1.

Corollary 4. Let H and H' be two hypergraphs. Then

$$\rho(H \times H') \leq \rho(H)\,\rho(H') \ .$$

Furthermore, if H is balanced, then $\rho(H \times H') = \rho(H)\,\rho(H')$ for every H' .

Clearly, if H^* is the dual of H , then $\rho(H) = \tau(H^*)$. If H is balanced, then H^* is also balanced.

Thus, the result follows immediately from Proposition 1, Proposition 2 and Corollary 1.

Corollary 5. Let H and H' be two hypergraphs. Then

$$\beta(H\times H') \geq \beta(H)\, n(H') + \beta(H')\, n(H) - \beta(H)\,\beta(H')$$

Equality holds for every H' if and only if $\tau(H) = \tau^*(H)$.

We have

$$\beta(H\times H') = n(H\times H') - \tau(H\times H') \geq n(H)\,n(H') - \tau(H)\,\tau(H') = n(H)\,n(H')$$
$$- (n(H)-\beta(H))\,(n(H')-\beta(H')) = \beta(H)\,n(H') + \beta(H')\,n(H) - \beta(H)\,\beta(H')$$

The equality holds iff it holds in Theorem 1.

Theorem 2. Let H and H' be two hypergraphs. Then

$$\tau(H\times H') \geq \tau(H) + \tau(H') - 1$$

A hypergraph $H = (E_i \,/\, i \in I)$ satisfies $\tau(H\times H') = \tau(H) + \tau(H') - 1$ for every H' if and only if $\bigcap_{i\in I} E_i \neq \emptyset$.

1. Let H and H' be two hypergraphs on X and Y respectively. Let T_o be a minimum transversal of $H\times H'$, and for $x \in X$, let

$$\varphi_o(x) = |\{y \,/\, (x,y) \in T_o,\ y \in Y\}|$$

Clearly, φ_o is an s-covering of H for $s = \tau(H')$. Let $T_1 \subset X\times Y$ be obtained from T_o by removing exactly $s-1$ vertices, and let

$$\varphi_1(x) = |\{y \,/\, (x,y) \in T_1,\ y \in Y\}|$$

We have, for all edges E_i of H,

$$\varphi_1(E_i) \geq \varphi_o(E_i) - (s-1) \geq 1$$

Hence, φ_1 is a 1-covering of H, and therefore $\varphi_1(X) \geq \tau(H)$. Hence

(1) $$\tau(H\times H') = \varphi_o(X) = \varphi_1(X) + (s-1) \geq \tau(H) + \tau(H') - 1 .$$

2. Now, consider a hypergraph $H = (E_i \,/\, i \in I)$ such that $\bigcap E_i \neq \emptyset$. Then $\tau(H) = 1$.

Let $x_o \in \bigcap E_i$. Clearly, $H\times H'$ has a transversal $T_o \subset \{x_o\}\times Y$ such that

$$|T_o| = \tau(H') = \tau(H) + \tau(H') - 1$$

Hence, by part 1 of the theorem, T_o is a minimum transversal of $H\times H'$, and

$$\tau(H\times H') = |T_o| = \tau(H) + \tau(H') - 1$$

Since this equality holds for every H' , the second part of the theorem is proved.

3. It remains to show that if $\tau(H) > 1$, there exists a hypergraph H' such that $\tau(H \times H') > \tau(H) + \tau(H') - 1$. Take any balanced hypergraph H' with $\tau(H') = s \geq 2$. By Corollary 1 to Theorem 1, we have

$$\tau(H \times H') = s\,\tau(H) > \tau(H) + (s-1) = \tau(H) + \tau(H') - 1$$

The required inequality follows.

Q.E.D.

Remark. Proposition 2 shows that, for all p, q ,

(1) $$\max\{\tau(H \times H') \,/\, \tau(H) = p,\ \tau(H') = q\} = pq$$

However, Theorem 2 shows only that

(2) $$\min\{\tau(H \times H') \,/\, \tau(H) = p,\ \tau(H') = q\} = p + q - 1$$

holds for $p = 1$ (or $q = 1$). However, it is easy to show that (2) holds for all pq .

Put $H = K^q_{p+q-1}$, $H' = K^p_{p+q-1}$ (the complete hypergraphs on $p+q-1$ vertices with ranks respectively q and p). Clearly, $\tau(H) = p$, $\tau(H') = q$. If the vertex set of H is $\{x_1, \dots, x_{p+q-1}\}$ and the vertex set of H' is $\{y_1, \dots, y_{p+q-1}\}$, then $T_o = \{(x_1, y_1), (x_2, y_2), \dots, (x_{p+q-1}, y_{p+q-1})\}$ is a transversal of $H \times H'$ because otherwise there exists an edge E_i of H and an edge F_j of H' such that $(E_i \times F_j) \cap T_o = \emptyset$, which contradicts that $|E_i| + |F_j| = p + q$. Thus, (2) follows from Theorem 2.

In fact, we can have a better inequality by using the number τ^* . We have

Theorem 3. Let H and H' be two hypergraphs. Then

$$\tau(H \times H') \geq \max\{\tau^*(H)\,\tau(H'),\ \tau(H)\,\tau^*(H')\} .$$

Let T_o be a minimum transversal of $H \times H'$, and let

$$\varphi_o(x) = |\{y \,/\, (x, y) \in T_o,\ y \in Y\}|$$

φ_o is an s-covering of H for $s = \tau(H')$. Hence, by Lemma 1,

$$\tau(H \times H') = |T_o| = \varphi_o(X) \geq \tau_s(H) \geq s\,\tau^*(H) = \tau(H')\,\tau^*(H) \quad .$$

The required inequality follows.

Corollary 1. $\tau(H \times H') \geq \max\{\frac{m(H)}{\delta(H)}\,\tau(H')\,,\ \frac{m(H')}{\delta(H')}\,\tau(H)\}$

This follows immediately from Lemma 1.

Corollary 2. $\tau(H \times H') \geq \max\{\nu(H)\,\tau(H')\ ,\ \nu(H')\,\tau(H)\}$

This follows immediately from Lemma 1.

3. The Chromatic Number. We shall now consider the chromatic number of the direct product $H \times H'$.

Example. (Polarized partition relations among cardinal numbers, [6], [4]). What is the least number of colors required to color the points of an $m \times n$ rectangle unit lattice so that rs points situated in r columns and s rows cannot have the same color? Clearly, this number is $\chi(K_m^r \times K_n^s)$.

For instance, $\chi(K_5^2 \times K_4^2) = 2$, and a bicoloring of the 6×4 rectangle unit lattice is shown in Example 2, Section 2.

Also, we have

$$\chi(K_5^2 \times K_5^2) = 3$$

Otherwise, there exists a bicoloring of the 5×5 matrix $((a_j^i))$ where the 0's denote the points of the first color and the 1's the points of the second color. Since the first column $(a_1^1, a_1^2, a_1^3, a_1^4, a_1^5)$ necessarily has three entries of equal value , suppose $a_1^1 = a_1^2 = a_1^3 = 0$.

The first two rows have, in each column, one of the combinations 00 , 11 , 01 , 10 , and there exist two columns with the same combination (because $2^2 < 5$) .

Since this repeated combination cannot be 00 nor 11 , we may assume

$$a_2^1 = a_3^1 = 0$$

$$a_2^2 = a_3^2 = 1 .$$

None of a_2^3, a_3^3 can be zero; hence

$$a_2^3 = a_3^3 = 1 .$$

Since the submatrix

$$\left(\left(\begin{matrix} a_2^2 & a_3^2 \\ a_2^3 & a_3^3 \end{matrix}\right)\right)$$

has only ones , the 0's and 1's in $((a_j^i))$ do not define a bicoloring of $K_5^2 \times K_5^2$.

Q.E.D.

This argument has been extended by Chvátal [3], [4], who showed that

(A) $$c_1 n^{1/r} \leq \chi(K_n^r \times K_n^r) \leq c_2 n^{1/r}$$

In fact, the lower bound also follows from a result of Kövary, Sós, Turán [9], while the upper bound was obtained by so-called probabilistic methods. Moreover, replacing the probabilistic method by a finite geometrical construction, one can show that

(B) $$\chi(K_n^2 \times K_n^2) / n^{1/2} \to 1$$

Finally, Sterboul [11] showed that in some cases, the same kind of arguments gives the exact value of $\chi(K_m^2 \times K_n^2)$.

The problem of finding a lower bound for $\chi(H \times H')$ was also considered by Chvátal [3], who gave the two following **inequalities:**

$$\chi(H \times H') \geq \min \{\chi(H) , \chi(H')^{1/n(H)}\} ,$$
$$\chi(H \times H') \geq \min \{\chi(H) , m(H)^{-1} \chi(H')\} .$$

An obvious result is:

<u>Proposition 3</u>. $\chi(H \times H') \leq \min \{\chi(H) , \chi(H')\}$

Assume that $\chi(H) \leq \chi(H')$, and let $g(x)$ be a coloring of H in $p = \chi(H)$ colors. Then $h(x , y) = g(x)$ is a coloring of $H \times H'$ in p colors. Hence $\chi(H \times H') \leq \chi(H)$.

Q.E.D.

Equality is obtained in some degenerate cases, for example when $\chi(H) = 2$. However, in general, Proposition 3 is far from being best possible. A better estimation for $\chi(H \times H')$, knowing $\chi(H) = p$ and $\chi(H') = q$, is:

__Theorem 4.__ $\max\{\chi(H \times H') / \chi(H) = p,\ \chi(H') = q\} = \chi(K_p^2 \times K_q^2)$

We have only to show that if H and H' are two hypergraphs with $\chi(H) = p$, $\chi(H') = q$, then

$$\chi(H \times H') \leq \chi(K_p^2 \times K_q^2)$$

Consider a coloring $c(x)$ of H with p symbols $a_1, a_2, \dots, a_p$, and a coloring $c'(y)$ of H' with q symbols $b_1, b_2, \dots, b_q$. Consider a complete graph K_p^2 with vertex set $\{a_1, a_2, \dots, a_p\}$ and a complete graph K_q^2 with vertex set $\{b_1, b_2, \dots, b_q\}$. Let $g(a_i, b_j)$ be a coloring of $K_p^2 \times K_q^2$ in $t = \chi(K_p^2 \times K_q^2)$ colors. Now, put

$$h(x, y) = g(c(x), c'(y))$$

To show that $h(x, y)$ is a coloring of $H \times H'$, consider an edge $E \times F$ of $H \times H'$. E contains two vertices x_1 and x_2 with $c(x_1) \neq c'(x_2)$, and F contains two vertices y_1 and y_2 with $c'(y_1) \neq c'(y_2)$. Since $\{c(x_1), c(x_2)\} \times \{c'(y_1), c'(y_2)\}$ is an edge of $K_p^2 \times K_q^2$, it contains two points, say $(c(x_3), c'(y_3))$ and $(c(x_4), c'(y_4))$, with

$$g(c(x_3), c'(y_3)) \neq g(c(x_4), c'(y_4))$$

Hence, $E \times F$ contains two vertices (x_3, y_3) and (x_4, y_4) with $h(x_3, y_3) \neq h(x_4, y_4)$. This shows that $h(x, y)$ is a t-coloring of $H \times H'$. Hence $\chi(H \times H') \leq t = \chi(K_p^2 \times K_q^2)$.

Q.E.D.

The problem of finding a good estimate for

$$f(p, q) = \min\{\chi(H \times H') / \chi(H) = p,\ \chi(H') = q\}$$

seems to be difficult. In particular, we can ask if as p and q tend to infinity, $f(p, q)$ tends to infinity.

REFERENCES

1. Berge, C., Graphes et Hypergraphes, Dunod, Paris 1970.

2. Brown, W. G., On graphs which do not contain a Thomsen graph, Canad. Math. Bull. 9, 1966, 281-285.

3. Chvátal, V., Hypergraphs and Ramseyian theorems, Thesis, Univ. of Waterloo, 1970.

4. Chvátal, V., On finite polarized partition relations, Canad. Math. Bull. 12, 1969, 321-326.

5. Čulik, K., Teilweise Losung eines verallgemeinerten Problems von K. Zarankiewicz, Ann. Polon. Math. 3, 1956, 165-168.

6. Erdös, P., and Rado, R., A partition calculus in set theory, Bull. A.M.S. 62, 1956, 427-489.

7. Guy, R. K., A problem of Zarankiewicz in Theory of Graphs, Akadémiai Kiadó, Budapest, 1968, 119-150.

8. Guy, R. K., A many-facetted problem of Zarankiewicz, in The many Facets of Graph Theory, Lecture Notes 110, Springer Verlag, Berlin 1969, 129-148.

9. Kövary, T., Sós, V., and Turán, P., On a problem of Zarankiewicz, Colloq. Math. 3, 1954, 50-57.

10. Lovász, L., Minimax theorems for hypergraphs, in Hypergraph Seminar, Lecture Notes, Springer Verlag, Berlin 1974.

11. Sterboul, F., On the chromatic number of the direct product of two hypergraphs, this volume, p.173.

GRAPHE REPRESENTATIF DE L'HYPERGRAPHE h-PARTI COMPLET

Jean-Claude Bermond, C.N.R.S.

I. Introduction

Définition 1. h étant un entier positif et $n_1 \leq n_2 \ldots \leq n_h$ un h-uple d'entiers positifs, on désignera par $K^h_{n_1,n_2,\ldots n_h}$ l'hypergraphe h-parti complet formé en considérant h ensembles disjoints, X_k, avec $|X_k| = n_k$; les sommets de $K^h_{n_1,n_2\ldots n_h}$ sont les éléments de $X = \bigcup_k X_k$ et les arêtes sont les parties E_i de X, telles que $|E_i \cap X_k| = 1$ pour tout k $(1 \leq k \leq h)$.

Dans le cas où $n_k = n$ pour tout k, on notera cet hypergraphe $K^h_{h \times n}$. Ces hypergraphes constituent une généralisation du graphe biparti complet (cas $h=2$ de la définition) et sont étudiés dans [3] et [13].

Définition 2. H étant un hypergraphe simple, $L_{h-1}(H)$ désigne le graphe simple, dont les sommets représentent les arêtes de H et où deux sommets sont reliés si et seulement si les arêtes qu'ils représentent ont une intersection de cardinalité $\geqslant h-1$. Dans le cas $h=2$, $L_1(H)$ n'est autre que le graphe représentatif des arêtes de l'hypergraphe H ([4] p. 383).

Remarque 1: Aux arêtes d'un hypergraphe H on associe aussi un multigraphe $R(H)$, dont les sommets représentent les arêtes de H et où deux sommets sont reliés par k arêtes dans $R(H)$, si les arêtes de H qu'ils représentent ont une intersection de cardinalité k. Dans le cas des hypergraphes h-parti complets, la donnée de $L_{h-1}(H)$ détermine $R(H)$ et $L_1(H)$. En effet 2 arêtes de H ont une intersection de cardinalité k si et seulement si les points qui les représentent dans $L_{h-1}(H)$ sont à une distance $h-k$.

Remarque 2: Les arêtes d'un hypergraphe h-parti complet peuvent être considérées comme les bases d'un matroïde sur X. $L_{h-1}(K^h_{n_1\ldots n_h})$ n'est autre alors que le "basis graph" de ce matroïde. (Le "basis graph" d'un matroide est le graphe simple dont les sommets représentent les bases du matroide, deux sommets étant reliés si et seulement si les bases associées ont une différence symétrique de cardinalité 2). Les "basis-graph" ont été étudiés par Bondy [5], Cunningham [6] et Maurer [12].

Nous nous proposons ici de caractériser $L_{h-1}(H)$ dans le cas où H est un hypergraphe h-parti complet, problème évoqué dans un article de Grünbaum [9 p. 119] (qui appelle $L_{h-1}(H)$ le "h-1 interchange graph"). Nous généraliserons ainsi des résultats de Hoffman [10], Moon [14] et Shrikhande [16] pour le cas h=2 (graphe représentatif des arêtes du graphe biparti complet) ; des résultats de Laskar [11] et Aigner [1] pour le cas h=3 ($L_2(K^3_{3 \times n})$ est appelé alors "cubic lattice graph") et de Rao [15] pour les cas h=4 et 5. Le problème analogue concernant les hypergraphes h-complets (les arêtes sont les parties à h éléments d'un ensemble X) a été étudié dans le cas général par Dowling [7] auquel on renvoie pour la bibliographie.

II. Résultats.

Nous nous limiterons ici au cas où $n_k = n$ pour tout k, c'est à dire à l'étude de $L_{h-1}(K^h_{h \times n})$. La plupart des résultats peuvent se généraliser au cas où tous les n_k ne sont pas égaux, mais les conditions obtenues sont très lourdes.

Notations : Dans un graphe G on désignera par d(x,y) la distance entre deux points x et y (longueur d'une plus courte chaîne entre x et y), par $D_i(x)$ l'ensemble des points à distance i de x.

On a la proposition suivante immédiate.

Proposition : $L_{h-1}(K^h_{h \times n})$ vérifie les propriétés suivantes

P_1 : son nombre de sommets est n^h

P_2 : c'est un graphe connexe régulier de degré $|D_1(x)| = h(n-1)$

P_3 : si $d(x,y) = 1$ alors $|D_1(x) \cap D_1(y)| = n-2$.

P_4 : si $d(x,y) = 2$ alors $|D_1(x) \cap D_1(y)| = 2$.

P_4 bis : si $d(x,y) = 2$ alors $D_1(x) \cap D_1(y)$ est formé de 2 points non adjacents.

P_5 : si $d(x,y) = 2$ alors $|D_1(x) \cap D_3(y)| \geqslant (h-2)(n-1)$.

P_7 : si $d(x,y) = 3$ alors $|D_1(x) \cap D_2(y)| \leq 3$.

Remarque : $L_{h-1}(K^h_{h \times n})$ vérifie aussi la propriété P_6 introduite dans [1].

P_6 : pour tout sommet x, $D_1(x)$ est l'union de h cliques, à n-1 sommets, 2 à 2 disjointes.

Nous verrons plus loin (lemme 3) que P_4^{bis} et P_6 sont équivalentes pour un graphe qui vérifie P_2, P_3, P_4. Tout graphe qui vérifie P_4 et ne vérifie pas P_4^{bis} contient un sous graphe isomorphe à G_o, où G_o est le graphe complet à 4 sommets moins une arête.

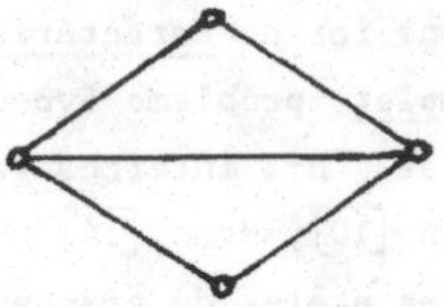

G_o

Le problème est de savoir dans quelle mesure ces propriétés caractérisent $L_{h-1}(K_h^h \times n)$. On a les résultats suivants.

Cas h=2.

Théorème 1 (Hoffman[10] et Moon [14]) : Si G est un graphe vérifiant P_1, P_2, P_3, P_4 pour $h=2$ et si $n \neq 4$, alors G est isomorphe à $L(K_2^2 \times n)$.

Théorème 2 (Shrikhande[16]) : Pour $h=2$ et $n=4$, il existe un graphe et un seul vérifiant P_1, P_2, P_3, P_4 et non isomorphe à $L(K_2^2 \times 4)$.

La matrice d'incidence de ce graphe est

	1	2	3	4	5	6	7	8	9	10	11	12	13	14	15	16
1	0	1	1	1	1	1	1	0	0	0	0	0	0	0	0	0
2		0	1	0	0	0	1	1	1	1	0	0	0	0	0	0
3			0	1	0	0	0	1	0	0	1	1	0	0	0	0
4				0	1	0	0	0	0	0	1	0	1	1	0	0
5					0	1	0	0	1	0	0	0	1	0	1	0
6						0	1	0	0	0	0	1	0	0	1	1
7							0	0	0	1	0	0	0	1	0	1
8								0	1	0	0	1	1	0	0	1
9									0	1	0	0	1	0	1	0
10										0	1	0	0	1	1	0
11											0	1	0	1	1	0
12												0	0	0	1	1
13													0	1	0	1
14														0	0	1
15															0	0
16																0

Remarque : On peut vérifier que le graphe exception, donné ci-dessus, ne vérifie pas la propriété P_4^{bis} ou P_6 : par exemple on a $d(3,7) = 2$ $D_1(3) \cap D_1(7) = \{1,2\}$ or 1 et 2 sont adjacents. On peut aussi remarquer que $D_1(1)$ est un cycle de longueur 6 et non l'union de 2 cliques à 3 sommets.

Pour une démonstration des théorèmes 1 et 2 on peut voir Aigner [2].

Cas h=3.

Théorème 3 (Laskar [11] et Aigner [1]) : Si G est un graphe vérifiant P_1, P_2, P_3, P_4, P_5 pour h=3 et si $n \neq 4$, alors G est isomorphe à $L_2(K_3^3 \times n)$.

Ce résultat a d'abord été prouvé par Laskar pour $n > 7$ et ensuite par Aigner pour tout $n \neq 4$ par une autre méthode. Aigner a aussi prouvé :

Théorème 4 : (Aigner [1]) : Si G est un graphe vérifiant P_1, P_2, P_3, P_4, P_5 et P_6 (ou P_4^{bis}) pour h=3, alors G est isomorphe à $L_2(K_3^3 \times n)$.

Théorème 5 : (Aigner [1]) : Pour h=3 et n=4 il existe un graphe et un seul vérifiant P_1, P_2, P_3, P_4, P_5 non isomorphe à $L_2(K_3^3 \times 4)$.

Théorème 6 : (Dowling [8]) : Si $n > 7$ et h=3, P_5 est conséquence de P_1, P_2, P_3, P_4.

Cas h=4 et h=5.

Théorème 7 : (Rao [15]) : Si G est un graphe vérifiant P_1, P_2, P_3, P_4 et P_5 pour h=4 ou 5 et si $n \geq 1 + \frac{h(h+1)}{2}$ alors G est isomorphe à $L_{h-1}(K_h^h \times n)$.

Nous nous proposons ici (en suivant la méthode d'Aigner [1]) de généraliser d'une part le théorème 4, puis les théorèmes 1, 3, 7 et enfin les théorèmes 2 et 5.

Théorème A : Si G est un graphe vérifiant P_1, P_2, P_3, P_4, P_4^{bis}, P_5 et P_7, pour h et n, alors G est isomorphe à $L_{h-1}(K_h^h \times n)$.

Théorème B : La propriété P_7 se déduit des propriétés P_2, P_3, P_4, P_4^{bis} et P_5 pour $h \leq 5$ et $n \geq h-1$.

Théorème C : La propriété P_4^{bis} se déduit des propriétés P_2 P_3 et P_4 pour $n \geq h(1+\sqrt{2})-1$.

En d'autres termes, si G est un graphe vérifiant P_1, P_2, P_3, P_4, P_5 et P_7, pour h et n, alors G est isomorphe à $L_{h-1}(K^h_{h \times n})$ si $n \geq h(1+\sqrt{2})-1$.

Théorème D : Pour $n=4$ et $h \geq 2$ il existe un graphe vérifiant P_1, P_2 P_3, P_4, P_5 et P_7 et non isomorphe à $L_{h-1}(K^h_{h \times n})$.

III. Démonstration des théorèmes A, B, C, D.

Définition. (voir [4] p. 363) - Etant donnés deux graphes simples $G_1 = (X,E)$ et $G_2 = (Y,F)$ on appelle somme cartésienne de G_1 et G_2 et on note G_1+G_2, le graphe dont l'ensemble des sommets est le produit cartésien des ensembles X et Y et où deux sommets (x,y) et (x',y') de $X \times Y$ sont reliés si et seulement si on a : soit $\{x = x'$ et $yy' \in F\}$, soit $\{xx' \in E$ et $y=y'\}$. (Cette opération est appelée "product" et notée $G_1 \times G_2$ dans [5] ou dans le livre d'Harary - Graph Theory - , Addison Wesley).

Remarque : Nous utiliserons ici la somme cartésienne de G et K_n (graphe complet à n sommets) ; $G+K_n$ peut être considéré comme un graphe formé de n copies de G (c'est à dire de n sous graphes isomorphes à G, $\varphi_i(G)$, $(i=1,2..n)$ tels que deux points de 2 copies différentes sont reliés si et seulement si ils sont images du même point de G).

La proposition suivante sera très utile : elle ramène le problème de la caractérisation de $L_{h-1}(K^h_{h \times n})$ à la caractérisation de la somme cartésienne de h graphes complets.

Proposition. $L_{h-1}(K^h_{n_1,n_2 \ldots n_h})$ est isomorphe à $K_{n_1}+K_{n_2}+\ldots+K_{n_h}$ (K_{n_k} désigne le graphe complet ou clique à n_k sommets).

Démonstration : On va montrer que $L_{h-1}(K^h_{n_1,n_2,\ldots n_h})$ est isomorphe à $L_{h-2}(K^{h-1}_{n_1,n_2 \ldots n_{h-1}}) + K_{n_h}$, ce qui par récurrence entrainera la proposition, $L_o(K_{n_1})$ étant isomorphe à K_{n_1}. Soient $x_1,x_2 \ldots x_{n_h}$ les éléments de X_h. Une arête de $K^h_{n_1,\ldots,n_h}$ s'écrit $E \cup \{x_i\}$ où E est une arête de $K^{h-1}_{n_1,\ldots,n_{h-1}}$. On peut donc écrire les sommets de $L_{h-1}(K^h_{n_1 \ldots n_h})$ sous la forme (e,i) où e est un sommet de $L_{h-2}(K^{h-1}_{n_1,\ldots n_{h-1}})$ et i un sommet de K_{n_h}. Deux sommets (e,i) et (e',i') de $L_{h-1}(K^h_{n_1 \ldots n_h})$ sont reliés si et seulement si les arêtes associées vérifient : $|\{E \cup \{x_i\}\} \cap \{E' \cup \{x_{i'}\}\}| = h-1$, donc si l'on a :

soit $E=E'$ et $x_i \neq x_{i'}$

soit $|E \cap E'| = h-2$ et $x_i = x_{i'}$.

Ceci revient à dire que (e,i) et (e',i') sont reliés si et seulement si :

soit $e=e'$ et ii' est une arête de K_{n_h}

soit ee' est une arête de $L_{h-2}(K^{h-1}_{n_1 \dots n_{h-1}})$ et $i=i'$ C.Q.F.D.

<u>Lemme 1</u>. G <u>étant un graphe simple il y a équivalence entre</u> :

(i) G <u>vérifie les propriétés</u> $P_1, P_2, P_3, P_4, P_5, P_7$ <u>aux ordres</u> n <u>et</u> h

(ii) $G+K_n$ <u>vérifie les propriétés</u> $P_1, P_2, P_3, P_4, P_5, P_7$ <u>aux ordres</u> n <u>et</u> h+1.

<u>Démonstration</u> : Nous noterons (x,i) un point de $G+K_n$, avec x sommet de G et i sommet de K_n. Remarquons que G est isomorphe aux sous graphes de $G+K_n$ formé des points (x,i_o) avec i_o fixé.

1. Le nombre de sommets de $G+K_n$ est égal à n fois le nombre de sommets de G d'où l'équivalence entre G vérifie P_1 aux ordres n et h et $G+K_n$ vérifie P_1 aux ordres n et h+1.

2. On a $D_1(x,i) = \{(z,i)$ où $z \in D_1(x)\} \cup \{(x,j)$ avec $j \neq i\}$ d'où $|D_1(x,i)| = |D_1(x)| + n-1$ et de plus $G+K_n$ est connexe si et seulement si G est connexe. On a donc l'équivalence entre G vérifie P_2 aux ordres n et h et $G+K_n$ vérifie P_2 aux ordres n et h+1.

3. Soient (x,i) et (y,j) deux points différents de $G+K_n$.

<u>Cas 1</u> : $i=j$ donc $x \neq y$

Dans ce cas $d((x,i), (y,i)) = d(x,y)$.

De plus $D_1(x,i) \cap D_1(y,i) = \{(z,i)$ avec $z \in D_1(x) \cap D_1(y)\}$. Si $d(x,y) = 2$:

$D_1(x,i) \cap D_3(y,i) = \{(z,i)$ avec $z \in D_1(x) \cap D_3(y)\} \cup \{(x,j)$ avec $j \neq i\}$.

Si $d(x,y) = 3$: $D_1(x,i) \cap D_2(y,i) = \{(z,i)$ avec $z \in D_1(x) \cap D_2(y)\}$.

Ces égalités montrent que si $G+K_n$ vérifie P_i $(i = 3,4,5,7)$ aux ordres n et h+1, G vérifie P_i $(i = 3,4,5,7)$ aux ordres n et h (fixer $i = i_o$) et que si G vérifie P_i $(i = 3,4,5,7)$ aux ordres n et h, les couples (x,i), (y,i) de $G+K_n$ vérifient P_i $(i = 3,4,5,7)$ aux ordres n et h+1.

Cas 2 : $i \neq j$

Dans ce cas $d((x,i)\ ,\ (y,j)) = d(x,y) + 1$.

. Si $d((x,i)\ ,\ (y,j)) = 1$, alors $x = y$ et :

$D_1(x,i) \cap D_1(x,j) = \{(x,k)$ avec $k \neq i,j\}$ soit $|D_1(x,i) \cap D_1(x,j)| = n-2$.

$G+K_n$ vérifie donc P_3 pour les couples $(x,i)\ ,\ (y,j)$ avec $i \neq j$.

. Si $d((x,i)\ ,\ (y,j)) = 2$, alors $d(x,y) = 1$ et :

$D_1(x,i) \cap D_1(y,j) = \{(x,j) \cup (y,i)\}$ donc $|D_1(x,i) \cap D_1(y,j)| = 2$.

$G+K_n$ vérifie donc P_4 pour les couples $(x,i)\ ,\ (y,j)$ avec $i \neq j$

. Si $d((x,i)\ ,\ (y,j)) = 2$ on a $2(n-1)$ points appartenant à $D_1(x,i) - D_3(y,j)$: les points (z,i) avec $z \in D_1(x) \cap D_1(y)$ en nombre $n-2$ si G vérifie P_2, les $(n-1)$ points (x,k) avec $k \neq i$ et le point (y,i). Si G vérifie P_1, alors $G+K_n$ vérifie P_1, donc $|D_1(x,i)| = (h+1)(n-1)$ et par suite $|D_1(x,i) \cap D_3(y,j)| = (h-1)(n-1)$. Donc si G vérifie P_1 et P_2, $G+K_n$ vérifie P_5 pour les couples $(x,i)\ ,\ (y,j)$ avec $i \neq j$.

. Si $d((x,i)\ ,\ (y,j)) = 3$ alors $d(x,y) = 2$ et :

$D_1(x,i) \cap D_2(y,j) = \{\ (z,i)$ avec $z \in D_1(x) \cap D_1(y)\} \cup \{x,j\}$, d'où si G vérifie P_4 $|D_1(x,i) \cap D_2(y,j)| = 3$. Donc si G vérifie P_4, $G+K_n$ vérifie P_7 pour les couples $(x,i)\ ,\ (y,j)$ avec $i \neq j$

C.Q.F.D.

Lemme 2. G étant un graphe simple il y a équivalence entre

(i) G vérifie P_4^{bis} aux ordres n et h.

(ii) $G+K_n$ vérifie P_4^{bis} aux ordres n et $h+1$.

Démonstration : On a vu au lemme 1 (3. Cas 1) que :

$D_1(x,i) \cap D_1(y,i) = \{(z,i)$ avec $z \in D_1(x) \cap D_1(y)\}$. Donc si $G+K_n$ vérifie P_4^{bis} , G vérifie P_4^{bis} et si G vérifie P_4^{bis} , les couples (x,i) (y,i) de $G+K_n$ vérifient P_4^{bis} . De plus on a vu dans le cas 2, que si $d((x,i),(y,j)) = 2$ et $i \neq j$ alors $D_1(x,i) \cap D_1(y,j) = \{(x,j) \cup (y,i)\}$; les deux points (x,j) et (y,i) sont dans ce cas non adjacents, d'où $G+K_n$ vérifie P_4^{bis} pour les couples (x,i) et (y,j) avec $i \neq j$.

C.Q.F.D.

Théorème D. Pour $n=4$ et $h \geq 2$ il existe un graphe vérifiant P_1, P_2, P_3, P_4, P_5, P_7 et non isomorphe à $L_{h-1}(K^h_{h \times n})$.

Démonstration : Pour $h=2$ un tel graphe existe d'après le théorème 2 (Skrikhande [16]) et on a vu qu'il ne vérifiait pas P_4^{bis}. On en déduit par récurrence, qu'il existe pour $n=4$ et $h \geq 2$ un graphe vérifiant P_1, P_2, P_3, P_4, P_5, P_7 (d'après le lemme 1 i $\Longrightarrow$ ii) et ne vérifiant pas P_4^{bis} (d'après le lemme 2 ii $\Longrightarrow$ i). Un tel graphe n'est pas isomorphe à $L_{h-1}(K^h_{h \times n})$, qui lui vérifie P_4^{bis}.

Notation : Dans la suite, lorsqu'aucune confusion ne sera possible, nous noterons $D_1(x)$ aussi bien l'ensemble des sommets adjacents à x, que le sous graphe engendré par cet ensemble (au lieu de $G_{D_1(x)}$)

La propriété P_6 s'énonce : (une clique étant un graphe complet).

P_6 : $D_1(x)$ est l'union de h cliques, à $n-1$ sommets, 2 à 2 disjointes (au sens des sommets).

Remarque : Si G vérifie de plus P_3, P_6 est équivalente à :

P_6^{bis} : $D_1(x)$ est formé de h composantes connexes, qui sont des cliques à $n-1$ éléments.

En effet, soit $y \in D_1(x)$, et soit C_y la clique à $n-1$ sommets contenant y d'après P_6. D'après P_3 $|D_1(x) \cap D_1(y)| = n-2$. Or $D_1(x) \cap D_1(y) \supset C_y - \{y\}$, donc $D_1(x) \cap D_1(y) = C_y - \{y\}$, d'où y est relié dans $D_1(x)$ aux seuls sommets de $C_y - \{y\}$ et C_y est bien une composante connexe.

Lemme 3. Soit G un graphe simple vérifiant P_2, P_3, P_4 alors les propriétés P_4^{bis} et P_6 sont équivalentes.

Démonstration : Supposons que G vérifie P_4^{bis} et soit x un sommet de G. Soit $z_o \in D_1(x)$, d'après P_3 on a $|D_1(x) \cap D_1(z_o)| = n-2$. Nous allons montrer que les $n-2$ points de $D_1(x) \cap D_1(z_o)$: $z_1 \ldots z_{n-2}$ forment une clique. En effet supposons que $d(z_i, z_j) \neq 1$, alors $d(z_i, z_j) = 2$; or $D_1(z_i) \cap D_1(z_j) \supset \{x, z_o\}$ avec x et z_o adjacents ce qui contredirait P_4^{bis}. Donc $z_o, z_1 \ldots z_{n-2}$ forment une clique à $n-1$ éléments dans $D_1(x)$ et de plus un sommet z_i de cette clique n'est relié à aucun autre point de $D_1(x)$ d'après P_3 appliquée à x et z_i. Enfin d'après P_2 $|D_1(x)| = h(n-1)$ ce qui prouve que $D_1(x)$ est bien formé de h cliques, à $n-1$ sommets, 2 à 2 disjointes.

Réciproquement, supposons que G vérifie P_6 et soient x et y deux sommets tels que $d(x,y) = 2$. Soient, d'après P_4, z_1 et z_2 les deux sommets de $D_1(x) \cap D_1(y)$. Supposons que z_1 et z_2 soient adjacents, alors $D_1(z_1) \supset \{x,y,z_2\}$. D'après P_6^{bis} (équivalente à P_6 d'après la remarque ci-dessus) x et z_2 étant adjacents doivent appartenir à une même clique de $D_1(z_1)$, de même y et z_2 appartiennent à une même clique de $D_1(z_1)$; mais x et y non adjacents appartiennent à 2 cliques différentes d'où une contradiction, ce qui prouve que z_1 et z_2 sont non adjacents.

C.Q.F.D.

Théorème C. La propriété P_4^{bis} (ou P_6) se déduit des propriétés P_2, P_3, P_4 pour $n \geq h(1+\sqrt{2})-1$.

Démonstration : De P_2, P_3, P_4 on déduit que le graphe engendré par $D_1(x)$ vérifie, pour tout x, les propriétés suivantes :

Q_1. son nombre de sommets est $h(n-1)$.

Q_2. Il est régulier de degré $n-2$.

Q_3. Si $d(x,y) = 2$ alors $|D_1(x) \cap D_1(y)| \leq 1$.

On est donc amené à étudier les graphes H vérifiant les propriétés Q_1, Q_2, Q_3 aux ordres h et n. Montrer le théorème C revient à montrer, que H est formé de h cliques, à $n-1$ sommets, 2 à 2 disjoints pour $n \geq h(1+\sqrt{2})-1$. Il suffit de montrer que pour ces valeurs de n, H admet une clique à $n-1$ sommets. En effet le graphe H^* obtenu en enlevant cette clique de H, vérifie les propriétés Q_1, Q_2, Q_3 aux ordres $h-1$ et n, et l'on a bien $n \geq (h-1)(1+\sqrt{2})-1$. En raisonnant par récurrence sur h, la propriété étant évidente pour $h=1$ (H étant alors la clique à $n-1$ sommets), H^* est l'union de $h-1$ cliques, à $n-1$ sommets, 2 à 2 disjointes et H est bien l'union de h cliques, à $n-1$ sommets, 2 à 2 disjointes.

Démonstration de : Pour $n \geq (1+\sqrt{2})-1$ H admet une clique à $n-1$ sommets.

1.- Soit C une clique maximale de H, alors si $x \notin C$, x est adjacent à au plus un point de C (immédiat d'après Q_3).

2.- Deux cliques maximales distinctes C et C' de H s'intersectent en au plus 1 point : ceci résulte de 1. appliqué à la clique C et à un point $x \notin C$ et $x \in C'$.

3.- Soit $x_i \in H, D_1(x_i)$ est l'union de cliques maximales C_{i_α} vérifiant

i). $C_{i_\alpha} \cup x_i$ est une clique maximale de H.

ii). Les C_{i_α} sont 2 à 2 disjointes (d'après 2 et 3. i)).

iii). Il n'y a pas d'arête entre C_{i_α} et C_{i_β} ($\alpha \neq \beta$) : en effet un point quelconque de C_{i_α} étant relié à x_i n'est relié à aucun autre point de $C_{i_\beta} \cup x_i$ d'après 1.

4.- <u>Dans la suite</u> C <u>désigne une clique de cardinalité maximum</u> ω : $C = \{x_1, x_2, \ldots, x_\omega\}$. Si $\omega = n-1$ le théorème est démontré, on suppose donc $\omega < n-1$ et <u>on pose $C_i = D_1(x_i) - C$</u>. D'après 3 C_i est l'union de cliques maximales vérifiant les propriétés 3 i). ii). iii).

5.- $|C_i| = n-\omega-1$ pour $i = 1,2,\ldots\omega$ (d'après Q_2).

6.- i). $C_i \cap C_j = \emptyset$ d'après 1 appliquée à C.

ii). Il n'y a pas d'arête entre C_i et C_j : en effet soit y un point de C_i ; $d(x_j, y) = 2$, donc d'après Q_3 $|D_1(x_j) \cap D_1(y)| \leq 1$ or $D_1(x_j) \cap D_1(y) \supset x_i$ donc y n'est relié à aucun autre point de C_j.

7.- Soit $y \in C_i$, alors $y \in C_{i_\alpha}$ (pour un certain α) et on a $|D_1(y) - \{C \cup D_1(C)\}| = n-2 - |C_{i_\alpha}|$: en effet si $z \in C \cup D_1(C)$ et $z \notin C_{i_\alpha} \cup x_i$, alors z n'est pas relié à y d'après 3 iii)(si $z \in C \cup C_i$) ou 6 ii). si $z \in C_j (j \neq i)$.

Donc $D_1(y) \cap \{C \cup D_1(C)\} = \{C_{i_\alpha} \cup \{x_i\}\} - \{y\}$, d'où le résultat compte-tenu de $|D_1(y)| = n-2$ d'après Q_2.

8.- Si $y \in C_i$ et $z \in C_i$ alors $\{D_1(y) \cap D_1(z)\} - \{C \cup D_1(C)\} = \emptyset$.

<u>Cas 1</u>. $y \in C_{i_\alpha}$ et $z \in C_{i_\alpha}$ la propriété résulte de 1 appliquée à $C_{i_\alpha} \cup x_i$.

<u>Cas 2</u>. $y \in C_{i_\alpha}$ et $z \in C_{i_\beta}$, alors $d(y,z) = 2$ d'après 3.iii) donc d'après Q_3 $|D_1(y) \cap D_1(z)| \leq 1$, or $x_i \in D_1(y) \cap D_1(z)$ d'où le résultat.

9.- Si $y \in C_i$ et $z \in C_j$ alors $|D_1(y) \cap D_1(z)| \leq 1$. Ceci résulte du fait que y et z sont non adjacents d'après 6 ii) et de Q_3.

10.- Première minoration du nombre de sommets de H.

Elle repose sur le fait que $C_{i_\alpha} \cup x_i$ étant une clique maximale $|C_{i_\alpha}| \leq \omega - 1$.

H contient au moins les sommets suivants :

. ceux de C en nombre ω

. ceux des C_i soit : $\sum_{i=1}^{\omega} |C_i| = \omega(n-\omega-1)$ d'après 5 et 6 i).

. ceux de $D_1(C_1)$ non encore rencontrés soit d'après 7 et 8. :

$\sum_\alpha |C_{1_\alpha}|\ (n-2-|C_{1_\alpha}|)$; compte-tenu de $|C_{1_\alpha}| \leq \omega - 1$ et de $\sum_\alpha |C_{1_\alpha}| = |C_1| = n-\omega-1$, il y a au moins $(n-\omega-1)^2$ tels sommets.

H contient donc au moins $\omega+\omega\,(n-\omega-1) + (n-\omega-1)^2$ sommets.

11.- Si $n > h+1$ alors $\omega \geq n-h$.

D'après Q_1, H a $h(n-1)$ sommets on doit donc avoir $\omega+\omega(n-\omega-1) + (n-\omega-1)^2 \leq h(n-1)$, soit $(n-\omega-1)\,(n-2) \leq (h-1)(n-1)$ ou encore $f(\omega) = (\omega+1-n)(n-2) + (h-1)(n-1) \geq 0$; $f(\omega)$ est une fonction croissante de ω et $f(n-h-1) = h+1-n$; donc si $n > h+1$, $f(n-h-1) < 0$ et par suite $f(\omega) < 0$ pour $\omega < n-h-1$. Donc si $n > h+1$, alors $\omega \geq n-h$.

12.- Deuxième minoration du nombre de sommets de H. Elle repose sur le fait que $|C_{i_\alpha}| \leq |C_i| = n-\omega-1$. Donc 7 s'écrit $|D_1(y) - \{C \cup D_1(C)\}| \geq \omega-1$.

H contient au moins les sommets suivants :

. ceux de C en nombre ω

. ceux de $\bigcup_i C_i$ soit : $\omega(n-\omega-1)$

. ceux de $D_1(C_1)$ non encore rencontrés, en nombre supérieur ou égal d'après 7 et 8 à $|C_1|(\omega-1) = (n-\omega-1)(\omega-1)$

. ceux de $D_1(C_2)$ non encore rencontrés, en nombre supérieur ou égal, d'après 7, 8, 9. à $|C_2|(\omega-1-|C_1|)$.

. et ainsi de suite les points de $D_1(C_i)$ non encore rencontrés en nombre supérieur ou égal à : $|C_i|\,(\omega-1-|C_1|-|C_2|-\dots-|C_{i-1}|)$ si cette quantité est positive, sinon 0.

H a donc, en posant $\lambda = i-1$, un nombre de sommets supérieur ou égal à :

$H(\omega) = \omega+\omega\,(n-\omega-1) + (\omega-1)(n-\omega-1)+\dots+(\omega-1)(\omega-1-\lambda(n-\omega-1))+\dots$

la sommation s'arrêtant au plus grand entier λ tel que

$\omega-1-\lambda(n-\omega-1) \geq 0$, soit à $\lambda_0 = \left[\frac{\omega-1}{n-\omega-1}\right]$ (on désigne par $[x]$ le plus grand entier inférieur ou égal à x). On doit avoir $i \leq \omega$ soit $\lambda \leq \omega-1$, ceci est bien réalisé car $\omega<n-1$ entraîne $\lambda_0 \leq \omega-1$.

H a donc au moins $H(\omega)$ sommets :

$$H(\omega) = \omega+\omega\,(n-\omega-1) + (n-\omega-1)\left(\sum_{\lambda=0}^{\lambda_0} \omega-1-\lambda(n-\omega-1)\right) \text{ où } \lambda_0 = \left[\frac{\omega-1}{n-\omega-1}\right].$$

13.- Si $n \geqslant 2h+1$ alors $\omega > n-h$.

On va montrer que $\omega = n-h$ conduit à une contradiction. En effet si $n \geqslant 2h+1$, alors $\lambda_0 \geqslant 1$ donc : $H(\omega) > \omega + \omega(n-\omega-1) + (\omega-1)(n-\omega-1)$. On doit avoir $H(\omega) \leqslant h(n-1)$. Or $H(n-h)-h(n-1) > (h-1)(n-2h-1)$. Donc si $n \geqslant 2h+1$, $H(n-h) > h(n-1)$ d'où la contradiction.

14.- Fin de la démonstration.

$$H(\omega) = \omega+\omega\,(n-\omega-1) + (n-\omega-1)\left((\lambda_0+1)(\omega-1) - \frac{\lambda_0(\lambda_0+1)}{2}(n-\omega-1)\right).$$ On a

$\lambda_0(n-\omega-1) = \left[\frac{\omega-1}{n-\omega-1}\right](n-\omega-1) \leqslant \omega-1$ d'où $H(\omega) \geqslant \omega+\omega\,(n-\omega-1)+(n-\omega-1)(\omega-1)\,\frac{\lambda_0+1}{2}$

Or $(\lambda_0+1)(n-\omega-1) > \omega-1$ soit $(\lambda_0+1)(n-\omega-1) \geqslant \omega$.

Donc $H(\omega) \geqslant \omega+\omega\,(n-\omega-1) + \frac{\omega(\omega-1)}{2}$. On doit avoir $H(\omega) \leqslant h(n-1)$ on est donc amené à étudier $F(\omega) = \omega^2-(2n-1)\omega + 2h(n-1)$ qui doit être $\geqslant 0$. Dans l'intervalle $[n-h+1, n-2]$ auquel appartient ω d'après 4 et 13 $F(\omega)$ est décroissante, donc $F(\omega) \leqslant F(n-h+1)$ où $F(n-h+1) = -n^2 + (2h+1)n + h^2 - 5h+2$.

On a un polynôme du 2ème degré en n dont le discriminant $\Delta = 8h^2 - 16h + 9 < (2h\sqrt{2}-3)^2$ donc $F(n-h+1) < 0$ si

$n \geqslant \frac{2h+1 + 2h\sqrt{2}-3}{2} = h\,(1+\sqrt{2})-1$. Pour $n \geqslant h(1+\sqrt{2})-1$ on a donc $F(\omega) < 0$ d'où une contradiction et alors $\omega = n-1$.

C.Q.F.D.

Remarque : La valeur trouvée donne que le théorème B est vrai pour $h = 3$ si $n \geqslant 7$, pour $h = 4$ si $n \geqslant 9$, pour $h = 5$ si $n \geqslant 12$. Pour $h = 5$ si on calcule directement $f(\omega)$ pour $n = 11$ on trouve $f(6) = 54$ $f(7) = 55$ $f(8) = 56$ $f(9) = 54$ et $h(n-1) = 50$. Donc dans ce cas $\omega = 10$ et pour $h = 5$ le théorème B est vrai pour $n = 11$. On peut montrer de même que pour $h = 6,7,8$ le théorème B est vrai pour $n \geqslant 2h+1$. On peut donc par un calcul plus poussé, espérer peut être, établir le théorème B pour $n \geq 2h+1$ (pour tout h). De toute façon par la méthode ci-dessus on ne peut espérer trouver mieux car pour $n=2h$ il existe des graphes H vérifiant Q_1, Q_2, Q_3 non formés de h cliques à $n-1$ sommets. Il faut donc recourir à des considérations directes comme l'a fait Aigner [1] pour $h=3$.

Dans la suite G désigne un graphe simple vérifiant P_2, P_3, P_4, P_4^{bis} et P_5 pour h et n. Pour démontrer les théorèmes A et B nous aurons besoin des définitions suivantes.

On appelle pont (bridge dans [15]) entre deux ensembles C et C' de sommets de G , une arête xy de G avec $x \in C-C'$ et $y \in C'-C$.

Deux cliques maximales distinctes C et C' de G seront dites couplées, s'il existe une bijection entre les sommets de C et ceux de C', telle qu'un sommet de C et un sommet de C' sont adjacents si et seulement si ils se correspondent dans la bijection ; deux tels sommets seront dits couplés. On peut dire aussi que $C \cup C' = C+K_2 = C'+K_2$. (Deux cliques couplées sont deux cliques vérifiant "$\delta(C,C') = 2$" dans [1] ou "appartenant à $\eta(G)$" dans [15]).

Remarquons que toute clique maximale de G contient n sommets d'après P_6 et que deux cliques couplées sont disjointes : en effet si $x \in C \cap C'$, il n'y a pas de pont entre C et C' (d'après P_6^{bis}). Si x et y sont adjacents, d'après P_6 il existe une unique clique maximale de G contenant x et y, que nous noterons $C_{x,y}$.

Lemme 4.- Soit G un graphe simple vérifiant P_2, P_3, P_4, P_4^{bis} et P_5. S'il existe deux ponts disjoints entre deux cliques maximales C et C' de G, alors les cliques C et C' sont couplées.

Démonstration : Soient x_1y_1 et x_2y_2 les deux ponts entre C et C' où x_1 et x_2 appartiennent à $C-C'$ $(x_1 \neq x_2)$ et y_1 et y_2 appartiennent à $C'-C$ $(y_1 \neq y_2)$. D'après P_6^{bis} les cliques $C-\{x_1\}$ et $C_{x_1y_1} - \{x_1\}$ sont 2 cliques disjointes de $D_1(x_1)$ sans pont entre elles ; donc si $x \in C$, $d(x,y_1) = 2$. A tout point x de C, $x \neq x_1$, associons le point $y = \varphi(x) = \{D_1(x) \cap D_1(y_1)\} - \{x_1\}$. D'après P_5 $|D_1(y_1) - D_3(x)| \leq 2(n-1)$. Or les $2(n-1)$ points des cliques $C_{y_1,y} - \{y_1\}$ et $C_{x_1,y_1} - \{y_1\}$ appartiennent à $D_1(y_1) - D_3(x)$, et $y_2 \in D_1(y_1) - D_3(x)$; $y_2 \notin D_1(x_1)$ donc $y_2 \in C_{y_1,y} - \{y_1\}$, d'où $C_{y_1,y} = C_{y_1,y_2} = C'$.
En posant $\varphi(x_1) = \{y_1\}$ on définit ainsi une application φ de C dans C'. Cette application est injective : en effet supposons que $y = \varphi(x) = \varphi(x')$ avec $x \neq x'$, y est différent de y_1, car $\varphi(x) \in C'-\{y_1\}$ pour $x \neq x_1$. Mais alors $D_1(x_1) \cap D_1(y) \supset \{x,x',y_1\}$ contrairement à P_4 appliquée à x_1 et y. L'application φ est bijective car $|C| = |C'| = n$.

C.Q.F.D.

Lemme 5. Soit G un graphe simple vérifiant P_2, P_3, P_4, P_4^{bis} et P_5. Soit $C = \{x_1, \ldots x_i, \ldots x_n\}$ une clique maximale de G contenant x_1 et soit $y_1 \in D_1(x_1)-C$, alors il existe une unique clique maximale couplée avec C et contenant y_1, qu'on notera $C_y = \{y_1, \ldots y_i, \ldots y_n\}$ où y_i désigne le point couplé avec x_i.

Démonstration : Soit $x_i \in C$, $x_i \neq x_1$. Toute clique couplée avec C et contenant y_1 doit contenir l'unique point (d'après P_4^{bis}) $y_i = \{D_1(x_i) \cap D_1(y_1)\} - \{x_i\}$. Or il existe une unique clique contenant y_1 et y_i : C_{y_1,y_i} et d'après le lemme 4 elle est couplée avec C (les deux ponts disjoints étant x_1y_1 et x_iy_i). C.Q.F.D.

Lemme 6. Soit G un graphe simple vérifiant P_1, P_2, P_3, P_4, P_4^{bis}, P_5 et P_7 aux ordres n et h, alors G peut s'écrire $G^* + K_n$.

Démonstration : On va montrer que G est formé de n copies $\varphi_i(G^*)$ d'un graphe G^*, que l'on va construire. Soit x_1 un point de G et soit $C = \{x_1, \ldots x_i, \ldots x_n\}$ une clique maximale contenant x_1. Posons $\varphi_i(x_1) = x_i$ pour $i = 1,2,\ldots n$.

1.- $D_1(C)$ est formé de n copies de $\{x_1\} \cup \{D_1(x_1)-C\}$ qui sont : $\{x_i\} \cup \{D_1(x_i)-C\}$ pour $i=1,2,\ldots,n$.

Soit $y_1 \in D_1(x_1)-C$ et soit $C_y = \{y_1, \ldots y_i, \ldots y_n\}$ l'unique clique couplée avec C contenant y_1 (lemme 5). Posons $\varphi_i(y_1) = y_i$, où y_i est le point couplé avec x_i ; y_i appartient donc à $D_1(x_i)-C$. Les sous graphes $D_1(x_i)-C$ sont d'après P_6^{bis} tous formés de $h-1$ composantes connexes, qui sont des cliques à $n-1$ éléments. Deux points de deux sous graphes $D_1(x_i)-C$ et $D_1(x_j)-C$ sont adjacents si et seulement si ils sont images par φ_i et φ_j du même point de $D_1(x_1)-C$: en effet si deux points $y_i = \varphi_i(y_1)$ et $y_j' = \varphi_j(y_1')$ étaient adjacents, avec $y_1 \neq y_1'$, il passerait par $\varphi_i(y_1) = y_i$ deux cliques couplées avec C à savoir : C_y et $C_{y_i,y_j'}$, ce qui est impossible d'après le lemme 5. De plus un point x_i n'est relié à aucun point de $D_1(x_j)-C$ (avec $j\neq i$) d'après la définition de deux cliques couplées. Si on prend $j=1$, on en déduit que l'application φ_i de $\{x_1\} \cup \{D_1(x_1)-C\}$ dans $\{x_i\} \cup \{D_1(x_i)-C\}$ est injective et bijective (les deux sous graphes ayant même nombre d'éléments). Il reste à vérifier que φ_i est bien un isomorphisme de graphes (c'est à dire conserve l'adjacence) : ceci résulte de ce que la clique C_{x_i,y_i} est couplée avec la clique $C_{x_1y_1}$ d'après le lemme 4 (on a deux ponts disjoints x_1x_i et y_1y_i). φ_i transforme donc toute clique maximale de $D_1(x_1)-C$ en une clique maximale de $D_1(x_i)-C$, de manière bijective.

2. $D_1(C) \cup D_2(C)$ est formé de n copies de $\{x_1\} \cup \{D_1(x_1)-C\} \cup \{D_2(x_1)-D_1(C)\}$

Soit $t_1 \in D_2(x_1)-D_1(C)$. D'après P_4^{bis} $t_1 = \{D_1(y_1) \cap D_1(z_1)\} - \{x_1\}$ avec y_1 et z_1 appartenant à $D_1(x_1)-C$ et $d(y_1,z_1) = 2$. Posons :

$\varphi_i(t_1) = t_i = \{D_1(y_i) \cap D_1(z_i)\} - \{x_i\}$ où $y_i = \varphi_i(y_1)$ et $z_i = \varphi_i(z_1)$ ont été définis au 1.

On a : $\underline{D_1(t_i) \cap D_1(C) = \{y_i\} \cup \{z_i\}}$: en effet t_i n'est pas adjacent à x_i (d'après P_4^{bis}), ni à un $x_j \neq x_i$ (d'après P_4 appliquée à x_i et t_i), ni à un point de $D_1(x_i)$ différent de y_i et z_i (d'après P_4 appliquée à x_i et t_i). Enfin t_i n'est pas adjacent à un point u_j de $D_1(x_j)$, car sinon $D_1(t_i)-D_3(x_i)$ contiendrait au moins $2(n-1)+1$ points : les points de $C_{t_i,y_i} - \{t_i\}$, ceux de $C_{t_i,z_i} - \{t_i\}$ et u_j, contrairement à P_5.

– $\underline{D_1(t_i) \cap \{D_2(x_j) - D_1(C)\} = \{t_j\}}$: En effet $|D_1(t_i) \cap D_2(x_j)| \leq 3$ d'après P_7 appliquée à t_i et x_j. $D_1(t_i) \cap D_2(x_j)$ contient les points de $D_1(t_i) \cap \{D_1(y_j) \cup D_1(z_j)\}$, soit y_i, z_i et les deux points : $\{D_1(t_i) \cap D_1(y_j)\} - \{y_i\}$ et $\{D_1(t_i) \cap D_1(z_j)\} - \{z_i\}$ qui sont donc confondus et identiques à $t_j = \{D_1(y_j) \cap D_1(z_j)\} - \{x_j\}$.

– En prenant $j=1$ on déduit que φ_i est une injection de $\{x_1\} \cup \{D_1(x_1)-C\} \cup \{D_2(x_1)-D_1(C)\}$ dans $\{x_i\} \cup \{D_1(x_i)-C\} \cup \{D_2(x_i)-D_1(C)\}$ et une bijection, les deux sous graphes ayant le même nombre d'éléments. Il reste à vérifier que φ_i est un isomorphisme de graphes.Or dans le sous graphe $\{x_1\} \cup \{D_1(x_1)-C\} \cup \{D_2(x_1)-D(C)\}$, t_1 est adjacent aux seuls points de $C_{y_1,t_1} - \{t_1\}$ et $C_{z_1,t_1} - \{t_1\}$; et dans la bijection φ_i, la clique C_{y_1,t_1} (C_{z_1,t_1}) est transformée en la clique C_{y_i,t_i} (C_{z_i,t_i}) qui lui est couplée d'après le lemme 4, car on a deux ponts disjoints $y_1y_i (z_1,z_i)$ et t_1t_i.

3. Supposons que l'on ait montré que pour $d \geq 2$, $\bigcup_{1 \leq \delta \leq d} D_\delta(C)$ est formé de n copies de $\{x_1\} \cup \{\bigcup_{1 \leq \delta \leq d} D_\delta(x_1)-D_{\delta-1}(C)\}$ (en posant $D_o(C) = C$). Soit $r_1 \in D_{d+1}(x_1)- D_d(C)$ et soit s_1 un point à distance 2 de r_1 sur une plus courte chaîne entre x_1 et r_1 et soient u_1 et v_1 les deux points de $D_1(s_1) \cap D_1(r_1)$. Si on pose $\varphi_i(r_1) = r_i = \{D_1(u_i) \cap D_1(v_i)\} - \{s_i\}$ où $u_i = \varphi_i(u_1)$; $v_i = \varphi_i(v_1)$ et $s_i = \varphi_i(v_1)$, d'après l'hypothèse de récurrence, les hypothèses du cas 2 sont remplies avec s_1 au lieu de x_1, u_1 au lieu de y_1, v_1 au lieu de z_1 et r_1 au lieu de t_1, on en déduit que φ_i réalise

une copie de $\{D_d(x_1)-D_{d-1}(C)\} \cup \{D_{d+1}(x_1) - D_d(C)\}$ sur

$\{D_d(x_i) - D_{d-1}(C)\} \cup \{D_{d+1}(x_i) - D_d(C)\}$. Comme de plus un point de $D_{d+1}(x_i)-D_d(C)$ n'est relié à aucun point de $\bigcup_{1 \leq \delta \leq d-1} D_\delta(C)$, sinon il serait à distance inférieure ou égale à d de x_i, on en déduit avec les hypothèses de récurrence que $\bigcup_{1 \leq \delta \leq d+1} D_\delta(C)$ est formé de n copies de $\{x_1\} \cup \{\bigcup_{1 \leq \delta \leq d+1} D_\delta(x_1) - D_{\delta-1}(C)\}$. Le graphe G étant connexe (d'après P_2) et fini (d'après P_1), $d_o = \max_{y \in G} d(x_1,y)$ est fini et G est formé de n copies de $G^* = \{x_1\} \cup \{\bigcup_{1 \leq \delta \leq d_o} D_\delta(x_1)-D_{\delta-1}(C)\}$. C.Q.F.D.

<u>Théorème A</u>. <u>Soit</u> G <u>un graphe simple vérifiant</u> P_1, P_2, P_3, P_4, P_4^{bis}, P_5 <u>et</u> P_7, <u>aux ordres</u> n <u>et</u> h, <u>alors</u> G <u>est isomorphe à</u> $L_{h-1}(K_h^h \times n)$.

<u>Démonstration</u> : On va démontrer le théorème par récurrence sur h. Le théorème est vrai pour $h=1$ (n quelconque), car on a un graphe à n sommets régulier de degré $n-1$ (d'après P_1 et P_2) c'est donc K_n. Supposons le théorème vrai aux ordres n et h, et soit G vérifiant les hypothèses aux ordres n et $h+1$. D'après le lemme 6 $G = G^* + K_n$. D'après les lemmes 1 et 2 (ii $\Longrightarrow$ i) G^* vérifie les hypothèses du théorème aux ordres n et h, donc d'après l'hypothèse de récurrence, G^* est isomorphe à $L_{h-1}(K_h^h \times n)$ et d'après la proposition 2, G est bien isomorphe à $L_h(K_{(h+1)}^{h+1} \times n)$. C.Q.F.D.

Des théorèmes A et C on déduit le corollaire.

<u>Corollaire</u>. <u>Soit</u> G <u>un graphe simple vérifiant</u> P_1, P_2, P_3, P_4, P_5 <u>et</u> P_7 <u>aux ordres</u> n <u>et</u> h, <u>avec</u> $n \geq h(1+\sqrt{2})-1$ <u>alors</u> G <u>est isomorphe à</u> $L_{h-1}(K_h^h \times n)$.

<u>Remarque</u> : Dans la démonstration du lemme 6 et du théorème A la propriété P_7 sert uniquement pour montrer dans le cas 2 du lemme 6 que t_i est adjacent à t_j et seulement à t_j dans $D_2(x_j) - D_1(C)$. Pour montrer le théorème B, nous montrerons que t_i est adjacent à t_j (lemme 9) indépendamment de P_7, ce qui est une étape dans la démonstration de la conjecture suivante : "P_7 se déduit de P_2, P_3, P_4, P_4^{bis} et P_5". Auparavant, nous montrons un lemme conséquence de P_5, qui sera très utile.

<u>Lemme 7</u>. <u>Soit</u> G <u>un graphe simple vérifiant</u> P_2, P_3, P_4, P_4^{bis}, <u>et</u> P_5 <u>et soient</u> C <u>et</u> C' <u>deux cliques maximales de</u> G. <u>S'il existe un pont</u> $x_1 y_1$ <u>entre</u> C <u>et</u> C' <u>et deux points</u> $x \in C$, $x \neq x_1$ <u>et</u> $z \in C'$, $z \neq y_1$, <u>tels que</u> $d(x,z)=2$, <u>alors</u> C' <u>est l'unique clique couplée avec</u> C, <u>contenant</u> y_1 (<u>soit</u> $C' = C_y$) <u>et</u> $D_1(x) \cap D_1(z) \subset C \cup C'$.

Démonstration : D'après le lemme 5, il existe une unique clique C_y contenant y_1 et couplée avec C. Supposons que C' soit différente de C_y ; elle est disjointe (d'après P_6) de C_y. Alors $D_1(y_1)-D_3(x)$ contiendrait les $2(n-1)$ points de $C_{x_1,y_1}-\{y_1\}$ et de $C_y-\{y_1\}$, plus le point z, ce qui est impossible d'après P_5. Donc $C'=C_y$. Comme $d(x,z) = 2$, $D_1(x) \cap D_1(z)$ est formé d'après P_4 de deux points, qui sont nécessairement le point de C' couplé avec x et le point de C couplé avec z, d'où $D_1(x) \cap D_1(z) \subset C \cup C'$.

C.Q.F.D.

Lemme 8. Soit G un graphe simple vérifiant P_2, P_3, P_4, P_4^{bis} et P_5 si $d(x,y) = 3$, alors $D_1(x) \cap D_2(y)$ est formé de points 2 à 2 non adjacents.

Démonstration : Soit x,a_1,b_1,y une chaîne de longueur 3 entre x et y. Supposons qu'il existe $a_2 \in D_1(x) \cap D_2(y)$ avec a_2 adjacent à a_1. D'après P_6^{bis}, a_2 appartient à la clique C_{x,a_1} et a_2 n'est pas adjacent à b_1 (les cliques $C_{a_1b_1}$ et C_{x,a_1} ayant deux points non adjacents x et b_1, n'ont que a_1 en commun). Soit $b_2 \in D_1(a_2) \cap D_1(y)$: un tel point existe d'après la définition de a_2, est différent de b_1 et de a_1 (sinon $d(x,y) = 2$). Les cliques C_{a_1,b_1} et C_{a_2,b_2} ont un pont a_1a_2 et deux points à distance 2 b_1 et b_2 (en effet, si b_2 était adjacent à b_1 les cliques C_{a_1,a_2} et C_{b_1,b_2} seraient couplées et alors $d(x,y)$ 2; d'après le lemme 7 y qui appartient à $D_1(b_1) \cap D_1(b_2)$ devrait appartenir à $C_{a_1b_1} \cup C_{a_2b_2}$ ce qui est impossible car y n'est pas adjacent à a_1 ni à a_2.

C.Q.F.D.

Corollaire. Si $h \leq 3$, la propriété P_7 se déduit des propriétés P_2, P_3, P_4, P_4^{bis} et P_5.

En effet $D_1(x)$ contient au plus h points 2 à 2 non adjacents d'après P_6^{bis}, d'où le résultat d'après le lemme 8.

Lemme 9. Soit G un graphe simple vérifiant P_2, P_3, P_4, P_4^{bis} et P_5 aux ordres n et h, avec $n \geq h-1$. Soit $C = \{x_1,\ldots x_i,\ldots x_n\}$ une clique maximale de G, soit y_1 et z_1 deux points de $D_1(x_1)-C$ non adjacents et soient $C_y = \{y_1,\ldots y_i,\ldots y_n\}$ et $C_z = \{z_1,\ldots z_i,\ldots z_n\}$ les cliques uniques couplées avec C_x contenant respectivement y_1 et z_1. Alors si on pose $t_i = \{D_1(y_i) \cap D_1(z_i)\} - \{x_i\}$, le sous graphe engendré par les $t_i (i=1,2\ldots n)$ est une clique C_t couplée avec C_y et C_z.

Démonstration : 1.- On a vu au 2 du lemme 6 (sans utiliser P_7) que : $D_1(t_i) \cap D_1(C) = \{y_i\} \cup \{z_i\}$.

2.- Soit C_{t_i} l'unique clique couplée avec C_y contenant t_i et soit α_i le point couplé avec y_1 ; $\alpha_i = C_{t_i} \cap D_1(y_1)$. Alors si $C_{t_i} \neq C_{t_j}$, $d(\alpha_i, \alpha_j) \geq 2$.

On peut supposer $j \neq 1$. α_i et α_j ne peuvent pas être confondus car sinon il passerait par α_i deux cliques couplées avec C_y, ce qui est impossible (lemme 5). Supposons que $d(\alpha_i, \alpha_j) = 1$. Alors les cliques C_{t_j} et C_{α_i, y_1} ont un pont $\alpha_i \alpha_j$ et deux points à distance 2, y_1 et t_j (car t_j n'est pas relié à y_1 d'après 1), donc d'après le lemme 7, y_j qui appartient à $D_1(y_1) \cap D_1(t_j)$ appartiendrait à $C_{t_j} \cup C_{\alpha_i, y_1}$ ce qui est impossible, car $y_j \notin C_{t_j}$ et y_j n'est pas relié à α_i (α_i étant couplé avec y_1).

3. <u>Si</u> $n \geq h-1$, <u>il existe deux indices</u> i_o <u>et</u> j_o <u>tels que</u> $C_{t_{i_o}} = C_{t_{j_o}}$.

En effet supposons que les C_{t_i} soient tous distincts, alors les α_i étant non adjacents (d'après 2) appartiendraient à des cliques disjointes de $D_1(y_1)$. De plus α_i n'appartient pas à C_y ni à C_{x_1, y_1} (d'après 1). On aurait donc $n+2$ cliques disjointes dans $D_1(y_1)$, en contradiction avec $n \geq h-1$ et P_6.

4. Soient i_o et j_o deux indices tels que $C_{t_i} = C_{t_j}$. Alors on a deux ponts disjoints $z_{i_o} t_{i_o}$ et $z_{j_o} t_{j_o}$ entre C_z et C_{t_i} donc d'après le lemme 4, C_z et C_{t_i} sont couplés. Soit s_k l'unique point de C_{t_i} couplé avec y_k. C_{t_i} étant couplé avec C_z, soit z_ρ le point couplé avec s_k. Supposons que $z_\rho \neq z_k$; alors les cliques C_{t_i} et C_{z_ρ, x_ρ} ont un pont $s_k z_\rho$ et deux points à distance 2 x_ρ et s_ρ point couplé avec y_ρ ; donc d'après le lemme 7, y_ρ, qui appartient à $D_1(x_\rho) \cap D_1(s_\rho)$, devrait appartenir à $C_{t_i} \cup C_{z_\rho, x_\rho}$ ce qui est impossible. Donc s_k est relié à z_k et $s_k = t_k$ ce qui prouve que t_k appartient à $C_{t_{io}}$, pour tout k.

C.Q.F.D.

<u>Remarque</u> : Dans le cas $h=4$ et $n=2$ on peut montrer que le lemme 9 est encore valable. On peut donc se demander si l'hypothèse $n \geq h-1$ est nécessaire pour montrer le lemme 9.

<u>Théorème B</u> : <u>La propriété</u> P_7 <u>se déduit des propriétés</u> P_2, P_3, P_4, P_4^{bis}, <u>et</u> P_5 <u>pour</u> $h \leq 5$ <u>et</u> $n \geq h-1$.

Démonstration : 1.- Soit G un graphe simple vérifiant P_2, P_3, P_4, P_4^{bis}, P_5, avec $n \geq h-1$ et soient x_o et y_o deux points tels que $d(x_o,y_o) = 3$ et soit x_o,a_1,b_1,y_o une chaîne de longueur 3 entre x_o et y_o. Posons $b_2 = \{D_1(a_1) \cap D_1(y_o)\} - \{b_1\}$; $a_2 = \{D_1(x_o) \cap D_1(b_1)\} - \{a_1\}$; $b_3 = \{D_1(a_2) \cap D_1(y_o)\} - \{b_1\}$; $a_3 = \{D_1(x_o) \cap D_1(b_2)\} - \{a_1\}$. Les b_i sont non adjacents à x_o et les a_i non adjacents à y_o, sinon $d(x_o,y_o) = 2$; ceci montre que les points a_1, a_2, a_3, b_1, b_2, b_3 sont tous distincts. (Remarquons qu'on a ainsi prouvé que si $d(x\ ,y\ ) = 3$ alors $|D_1(x\ ) \cap D_2(y)| \geq 3$, d'où P_7 est équivalente à P_7^{bis} : si $d(x,y) = 3$ alors $|D_1(x) \cap D_2(y)| = 3$).

2.- Montrons que a_3 est adjacent à b_3 : soit $t_1 = \{D_1(a_2) \cap D_1(a_3)\} - \{x_o\}$. Si on pose $C_x = C_{x_o,a_1}$; $C_y = C_{a_2,b_1}$ et $C_z = C_{a_3,b_2}$, les hypothèses du lemme 9 sont satisfaites, avec b_1 couplé avec a_1 et b_2 couplé avec a_1. On en déduit que t_1 est adjacent à $y_o = \{D_1(b_1) \cap D_1(b_2)\} - \{a_1\}$. On a donc $D_1(a_2) \cap D_1(y_o) \supset \{b_1,b_3,t_1\}$ comme $t_1 \neq b_1$ (d'après P_4 appliquée à x_o et b_1) on a $t_1 = b_3$ ce qui prouve que a_3 est adjacent à b_3.

3.- Supposons que P_7 soit non vrai c'est à dire qu'il existe $c_1 \in D_1(x_o) \cap D_2(y_o)$ avec $c_1 \neq a_i$ ($i = 1,2,3$) et soit x_o, c_1, d_1, y_o une chaîne de longueur 3 entre x_o et y_o. c_1 est différent de b_i et d_1 différent de a_i ($i = 1,2,3$) sinnn $d(x_o,y_o) = 2$. Enfin d_1 est différent de b_i ($i=1,2,3$) d'après P_4 appliquée à x_o et b_i. En recommençant le raisonnement de 1 avec la chaîne x_o, c_1, d_1, y_o on en déduit l'existence de c_2 et c_3 (ainsi que d_2 et d_3) tels que les c_i ($i = 1, 2, 3$) soient tous distincts et différents des a_j et b_j ($j = 1, 2, 3$). Donc, si P_7 est non vrai $|D_1(x_o) \cap D_2(y_o)| \geq 6$; d'après le lemme 8, les 6 points de $D_1(x_o) \cap D_2(y_o)$ sont 2 à 2 non adjacents, d'où $h \geq 6$.

C.Q.F.D.

En combinant les théorèmes A, B, C on a le corollaire.

Corollaire : P_1, P_2, P_3, P_4, P_5 caractérisent $L_{h-1}(K^h_{h \times n})$ pour $h=2$ et $n \geq 5$; $h=3$ et $n \geq 7$; $h=4$ et $n \geq 9$; $h = 5$ et $n \geq 11$.

Problèmes : Il serait intéressant de savoir dans quelle mesure les propriétés P_i sont indépendantes les unes des autres en particulier de savoir :

- si P_7 n'est pas conséquence des autres pour $n \geq h-1$ et pour n quelconque.

- si P_4^{bis} n'est pas conséquence de P_2, P_3, P_4 pour $n \geq 2h+1$ et pour $n \geq 5$.

- si P_5 n'est pas conséquence des autres P_i comme cela a été montré pour h = 3 par Dowling [8].

Note (ajoutée en cours de rédaction) : R. Laskar et A. Pellerin auraient prouvé (Notices A.M.S. n° 699-A. 18, Décembre 1972) que P_1, P_2, P_3, P_4 caractérisent $L_3(K^4_{4 \times n})$ si $n \geq 12$.

L'auteur remercie pour son aide précieuse Jean-Claude Meyer, sans qui cet article n'aurait pu être mené à bien.

REFERENCES

1. M. Aigner : The uniqueness of the cubic lattice graph. J. Combinatorial Theory 6 (1969), 282-297.
2. M. Aigner : Note on the characterization of certain association schemes. Ann. Math. Stat. 42 (1971), 363-367.
3. C. Berge : Nombres de coloration de l'hypergraphe h-parti complet, this volume.
4. C. Berge : Graphes et hypergraphes, Dunod, Paris, 1971.
5. J.A. Bondy : Pancyclic graphs II, J. Combinatorial Theory, (à paraître).
6. W. Cunningham : The matroids of a Basis Graph (à paraître).
7. T.A.Dowling : A characterization of the T_m graph. J. Combinatorial Theory 6 (1969), 251-263.
8. T.A. Dowling : Note on "a characterization of cubic lattice graphs". J. Combinatorial Theory 5 (1968), 425-426.
9. B. Grünbaum : Incidence patterns of graphs and complexes. The many facets of graph theory ; Springer-Verlag, Lecture Notes (110), 115-128.
10. A.J. Hoffman : On the line graph of the complete bipartite graph, Ann. Math. Stat. 35 (1964), 883-885.
11. R. Laskar : A characterization of cubic lattice graphs, J. Combinatorial Theory 3 (1967), 386-401.
12. S.B. Maurer : On matroid basis graphs. Notices Amer. Math. Soc., Août 1972, 617.
13. J.C. Meyer : Quelques problèmes concernant les cliques des hypergraphes h-complets et q-parti h complets, this volume.
14. J. W. Moon : On the line graph of the complete bigraph. Ann. Math. Stat. 34 (1963), 664-667.
15. A.R. Rao : A characterization of a class of regular graphs. J. Combinatorial Theory 10 (1971) 264-274.
16. S.S. Shrikhande : The uniqueness of the L_2 association scheme. Ann. Math. Stat. 30 (1959), 781-798.

THE CHROMATIC INDEX
OF AN INFINITE COMPLETE HYPERGRAPH :
A PARTITION THEOREM

R. Bonnet, University of Lyon,

P. Erdös, Hungarian Academy of Science

1. Introduction. Let S be an infinite set of power n, and let $m \geq 2$ be an integer. We denote by $P_m(S)$ or $[S]^m$ the set of all subsets of S, of power m. A corollary of the main theorem proves that : if we denote by K_n^m the complete hypergraph having $P_m(S)$ as a set of edges (so its degree is n), then K_n^m has n for chromatic index. This result gives a positive answer to a conjecture of C. Berge.

2. Notations. Subsequently we assume the axiom of choice : particularly every infinite cardinal is an initial ordinal, denoted by ω_α. Moreover ω_o is written ω. If S is a set, its cardinality is denoted by $|S|$. If $m < |S|$ is a cardinal, then $P_m(S)$ or $[S]^m$ is the set of all subsets Y of S so that $|Y| = m$: an element of $[S]^m$ is called a m-tuple.

3. Theorem 1. Let p, m and n be three cardinals so that $p < m < n$ and $1 \leq p < \omega \leq n$. If S is a set of power n, there exists a partition $(\Delta_k)_{k \in I}$ of $[S]^m$, with $|I| = n^m$, such that for every k in I, every p-tuple is included in exactly one m-tuple, member of Δ_k.

If $p = 1$, each Δ_k defines a partition of S : the distinct sets, members of Δ_k, are disjoint, and Δ_k is a covering of S. This solves a conjecture of C. Berge : for $2 \leq m < \omega$ and $|S| \geq \omega$, the complete hypergraph has a coloring of the edges such that each vertex meets all the colors.

From $1 \leq p < \omega$, it follows that for any p-tuple Z of S and any m-tuple A of S, the condition $Z \subset A$ is equivalent to : $|Z \cap A| = |Z| = p$. So this result is a corollary of the following theorem :

4. Theorem 2. Let p, m and n be three cardinals so that $1 \leq p < m < n$ and $n^p = n \geq \omega$. If S is a set of power n, there exists a partition $(\Delta_k)_{k \in I}$ of $[S]^m$, with $|I| = n^m$, so that for every $k \in I$, on the one hand for every p-tuple Z of S, there is a m-tuple A, member of Δ_k, so that $|Z \cap A| = p = |Z|$, on the other hand, for distinct members A' and A'' of Δ_k, we have $|A' \cap A''| < p$.

In this theorem, and contrary to what happens in theorem 1, whenever $p \geq \omega$, we cannot suppose that every p-tuple is included in exactly one member of Δ_k. In fact, if we consider a p-tuple Z and suppose there is a unique set A in Δ_k which contains Z, then let z be an element of $S - A$, so $|Z \cup \{z\}| = p$, and there is a set A', member of Δ_k, which contains $Z \cup \{z\}$. From $A' \neq A''$, we obtain a contradiction. Moreover, we cannot suppose that for every p-tuple Z in S, there is exactly one member A of Δ_k such that $|Z \cap A| = p = |Z|$: this is a consequence of the following remark : the union of two distinct p-tuples has p for power.

5. Proof of theorem 2. If such a partition exists, on one hand $n^p = n$, on the other hand for every distinct sets A' and A'' of Δ_k, we must have $|A' \cap A''| < p$. Therefore $|\Delta_k| = n$, and so $|I| = n^m$. Moreover $n^p = n = \omega_\eta$ and $n^m = \omega_\alpha$, so we denote by $(B_\xi)_{\xi < \omega_\alpha}$ an enumeration of $[S]^m$ and by $(Z_\lambda)_{\lambda < \omega_\eta}$ an enumeration of $[S]^p$. If m is finite, $m - p > 0$ is an integer, otherwise $m - p$ is the cardinal m.

Suppose we defined the family $(\Delta_k)_{k < \gamma}$, when $\gamma < \omega_\alpha$, in such a way that we have :

a. for $k' < k'' < \gamma$, the sets $\Delta_{k'}$ and $\Delta_{k''}$ are disjoint.

b. for $k < \gamma$, if A' and A'' are distinct sets of Δ_k then $|A' \cap A''| < p$.

c. for $k < \gamma$, and for every p-tuple Z in S, there is at least one set A, member of Δ_k, such that $|Z \cap A| = p = |Z|$.

d. if $\xi < \gamma$, for some $k < \gamma$, the m-tuple B_ξ is a member of Δ_k.

We will construct the family $(\Delta_{\gamma,\rho})_{\rho < \omega_\eta}$ of sets of m-tuples such that the union of this family is Δ_γ.

1. If B_γ is a set, member of an already constructed Δ_k, then $\Delta_{\gamma,0}$ is the empty set.

2. If B_γ belongs to no Δ_k, for $k < \gamma$, then $\Delta_{\gamma,o}$ is the singleton $\{B_\gamma\}$.

Let λ' be the smallest λ so that $|Z_{\lambda'} \cap A| < p$ for every set A, member of $\Delta_{\gamma,o}$ (we have $\lambda' = o$ iff : $\Delta_{\gamma,o}$ is empty ; or $|Z_o \cap B_\gamma| < p$). If $\Delta_{\gamma,o}$ is the empty set, we put $S_o = Z_{\lambda'}$, otherwise we put $S_o = Z_{\lambda'} \cup B_\gamma$. For every $k < \gamma$ there is at most one subset C_k of S, member of Δ_k so that $Z_{\lambda'} \subset C_k$. We know that $[S - S_o]^{m-p}$ has $n^{m-p} = n^m$ elements and thus there is a subset D of $S - S_o$ verifying $|D| = m-p$ and so that : for every $k < \gamma$, the set $D \cup Z_{\lambda'}$ which is a m-tuple, is not a member of Δ_k. We remark that for every A, member of $\Delta_{\gamma,o}$, we have $|(Z_{\lambda'} \cup D) \cap A| = |Z_{\lambda'} \cap A| < p$. Let ξ' be the smallest ξ in ω_α so that :

1. for $k < \gamma$, the set $B_{\xi'}$ does not belong to Δ_k.
2. we have $|Z_{\lambda'} \cap B_{\xi'}| = p$.
3. for every A in $\Delta_{\gamma,o}$, we have $|B_{\xi'} \cap A| < p$ (this is verified whenever $\Delta_{\gamma,o}$ is the empty set).

We put $\Delta_{\gamma,1} = \Delta_{\gamma,o} \cup \{B_{\xi'}\}$

Suppose we defined $(\Delta_{\gamma,\nu})_{\nu < \rho}$, when $1 \leq \rho < \omega_n$, and verifying the following properties :

i. for $\nu' < \nu'' < \rho$ the set $\Delta_{\gamma,\nu'}$ is included in $\Delta_{\gamma,\nu''}$

ii. for distinct m-tuples A' and A'' in $\Delta_{\gamma,\nu}$, then $|A' \cap A''| < p$

iii. if A is a set of $\Delta_{\gamma,\nu}$, for every $k < \gamma$, the set A does not belong to Δ_k.

If ρ is a limit ordinal, then $\Delta_{\gamma,\rho}$ is the union, for $\nu < \rho$, of $\Delta_{\gamma,\nu}$.

If ρ is an isolated ordinal, $\rho = \theta + 1$, let S'_1 be the union of all members A of $\Delta_{\gamma,\theta}$. From $|A| = m < n$ and $|\rho| < n$, it follows that $|S'_1| < n$. Let λ'' be the smallest λ so that for every set A, member of $\Delta_{\gamma,\theta}$, we have $|Z_{\lambda''} \cap A| < p$. Such a λ'' exists because $|S - S'_1| = n$. We put $S_1 = Z_{\lambda''} \cup S'_1$. For every $k < \gamma$ there is at most a set D_k, member of Δ_k, so that $Z_{\lambda''} \subset D_k$. We know that $[S - S_1]^{m-p}$ has $n^{m-p} = n^m$ elements, and thus there is a (m-p)-tuple G of $S - S_1$ so that : on the one hand $|Z_{\lambda''} \cup G| = m$ (obvious), on the other hand $Z_{\lambda''} \cup G$ does not belong to every already constructed Δ_k. Moreover for every set A, member of $\Delta_{\gamma,\theta}$, we have $|Z_{\lambda''} \cup G) \cap A| = |Z_{\lambda''} \cap A| < p$. So let ξ'' be the smallest ξ so that :

i. for any $k < \gamma$, the set $B_{\xi''}$ is not a member of Δ_k.

ii. we have $|Z_{\lambda''} \cap B_{\xi''}| = p$.

iii. for every member A of $\Delta_{\gamma,\theta}$, we have $|A \cap B_{\xi''}| < p$.

Hence, we put $\Delta_{\gamma,\rho} = \Delta_{\gamma,\theta} \cup \{B_{\xi''}\}$.

From the construction, it follows that the family $(\Delta_{\gamma,\nu})_{\nu \leqslant \rho}$ verifies the conditions (i) , (ii) and (iii) . Moreover if the family $(\Delta_{\gamma,\rho})_{\rho < \omega_\eta}$ is constructed, then we put $\Delta_\gamma = \bigcup_{\rho<\omega_\eta} \Delta_{\gamma,\rho}$.

So the family $(\Delta_k)_{k \leqslant \gamma}$ verifies (a), (b), (c) and (d) .

Our transfinite induction is complete, and so the family $(\Delta_k)_{k<\omega_\alpha}$ verifies the conclusion of theorem 2 .

6. The case when $n^p > n$. Now, we assume the general continuum hypothesis (g.c.h.) . Let n and p be two infinite cardinals such that $n^p > n$. From g.c.h., it follows (by well known theorems [3]) that $n = \omega_\eta$ is a singular cardinal and that its cofinal type $cf(\omega_\eta) = \omega_\beta$ verifies $cf(n) = \omega_\beta \leqslant p = \omega_\delta$. Moreover, if $n = \omega_\eta$, then $n^+ = \omega_{\eta+1}$.

6.1 Theorem 3. Let p , m and n be three infinite cardinals such that $\omega \leqslant p < m < n < n^p$, $cf(n) < p$ and $cf(n) \neq cf(p)$. If S is a set of power n , there exists a partition $(\Delta_k)_{k \in I}$ of $[S]^m$, with $|I| = n^m$, so that for every $k \in I$, on the one hand for every p-tuple Z of S , there is at least a m-tuple A , member of Δ_k , such that $|Z \cap A| = p = |Z|$ on the other hand, for distinct members A' and A'' of Δ_k , we have $|A' \cap A''| < p$.

Proof. Let S be a set of power n . So S is the union of an increasing family of sets $(S_\nu)_{\nu < \omega_\beta}$ such that : on one hand for $\nu' < \nu'' < \omega_\beta$ the set $S_{\nu'}$ is included in $S_{\nu''}$, on the other hand, for every $\nu < \omega_\beta$, we have $m < |S_\nu| = n_\nu < n$ and n_ν is a regular cardinal . From g.c.h., it follows that $[S_\nu]^p$ is a set of power n_ν . Let L be the union of $[S_\nu]^p$ for $\nu < \omega_\beta$, we have $|L| = n$. If we denote by $(Z'_\lambda)_{\lambda < \omega_\eta}$ an enumeration of L , by the method used in the proof of theorem 2 , we can construct a partition $(\Delta_k)_{k \in I}$ of $[S]^m$ such that for every k in I , on one hand, for every p-tuple Z' , member of L , there is a m-tuple A in Δ_k such that $|Z' \cap A| = p$, on the other hand, for distinct members A' and A'' of Δ_k , we have

$$|A' \cap A''| < p$$

<u>a</u>. p is a regular cardinal. Let Z be a p-tuple in S, there exists a $\nu < \omega_\beta$ so that $Z \cap S_\nu = Z'$ verifies $|Z'| = p$: since $\omega_\beta = cf(n) < p$ and $p = cf(p)$. Therefore, there is at least a member A of Δ_k so that $|Z' \cap A| = p$ (indeed Z' belongs to $[S_\nu]^p$) and so $|Z \cap A| = p$. From these remarks, it follows that the family $(\Delta_k)_{k \in I}$ satisfies the conclusions of the theorem.

<u>b</u>. p is a singular cardinal such that $cf(n) < cf(p) < p$. Let Z be a p-tuple in S, there is some ν so that $|Z \cap S_\nu| = p$: otherwise let Z_ν be the set $Z \cap S_\nu$; from $|Z_\nu| < p$ and

$$Z = \bigcup_{\nu < \omega_\beta} Z_\nu$$

it follows that $|Z| < p$ (this is a consequence of $\omega_\beta = cf(n) < cf(p)$). So, we conclude as before.

<u>c</u>. p is a singular cardinal such that $cf(p) < cf(n) < p$. If Z is a p-tuple in S, there is at least a ν such that $|Z \cap S_\nu| = p$. Otherwise let Z_ν be the set $Z \cap S_\nu$, so Z is the union of Z_ν for $\nu < \omega_\beta = cf(n)$ and we have $|Z_\nu| = p_\nu < p$. It follows that p is the lower upper bound of the family $(p_\nu)_{\nu < \omega_\beta}$. Since ω_β is a regular cardinal, we can suppose that we have $p_{\nu'} < p_{\nu''}$ for $\nu' < \nu'' < \omega_\beta$. Therefore, $cf(n) = cf(p)$, and we have a contradiction. We conclude as before.

<u>Remark</u>. Under theorem hypotheses, if L is a subset of $[S]^p$ such that for every distinct members Z' and Z'' of L, we have $|Z' \cap Z''| < p$, then $|L| \leqslant n$. Otherwise $|L| \geqslant n^+ = 2^n$ and for every Z, member of L, let $\nu(Z)$ be a $\nu < \omega_\beta$ such that $|Z \cap S_\nu| = p$. From $n^+ = 2^n$ is a regular cardinal and from $cf(n) = \omega_\beta < n < n^+$, it follows that for some ν_0 there are n^+ members Z of L such that $\nu(Z) = \nu_0$. Consequently there are at least two members Z' and Z'' of L such that $|S_{\nu_0} \cap Z' \cap Z''| = p$ (this is a consequence of $|[S_{\nu_0}]^p| = |S_{\nu_0}| < n^+$), and we obtain a contradiction. This result is due to Tarski [4].

6.2 <u>Theorem 4. Let p, m and n be three infinite cardinals such that $\omega \leqslant p < m < n < n^p$, and either $cf(n) = p$, or $cf(n) = cf(p)$. If S is a set of power n, there is no subset Δ of $[S]^m$ so that on one hand, for every p-tuple Z in S there is at least a member A of such that $|Z \cap A| = p$, on the other hand for distinct members A' and A'' of Δ, we have $|A' \cap A''| < p$.</u>

Consequently, there is no partition $(\Delta_k)_{k\in I}$ of $[S]^m$ such that every Δ_k verifies the properties of the Δ above. Frascella, in [2], uses some similar idea.

<u>Proof</u>. First, we will prove that if such a Δ exists, then $|\Delta| \geqslant n^+ = 2^n$. To show this, we suppose $|\Delta| \leqslant n$, and thus $|\Delta| = n = \omega_\eta$. We denote by $(A_\xi)_{\xi<n}$ an enumeration of all members of Δ. We know that $n = \omega_\eta$ is the union of $cf(n) = \omega_\beta$ strictly increasing sets n_ν for $\nu < \omega_\beta$, with $|n_\nu| < n$.

<u>a</u>. if we have $cf(n) = p$, then we can construct, by transfinite induction, a sequence of elements w_ν in S, such that w_ν does not belong to the union V_ν of A_ξ for $\xi \in n_\nu$, and w_ν is distinct from every already constructed $w_{\nu'}$. This is possible : indeed let W_ν be the union of V_ν and the set of $w_{\nu'}$ for $\nu' < \nu$, we have $|W_\nu| < n$ and thus $|S - W_\nu| = n$. We denote by Z the set of all w_ν for $\nu < p$.

<u>b</u>. if we have $cf(n) = cf(p) < p$, then p is a singular cardinal. Let $(p_\nu)_{\nu<cf(p)}$ be a partition of p so that $|p_\nu| < p$ for $\nu < cf(p)$. We construct, by transfinite induction, a sequence of subsets Z_ν of S, for $\nu < cf(p)$, such that : on one hand Z_ν is disjoint from every already constructed $Z_{\nu'}$, on the other hand Z_ν is disjoint from the union of A_ξ for $\xi \in n_\nu$. We denote by Z the union of Z_ν for $\nu < cf(p)$.

In these two cases Z verifies $|Z| = p$. From the construction of Z, it follows that for every member A_ξ of Δ, we have $|Z \cap A_\xi| < p$. Contradiction. So $|\Delta| \geqslant n^+ = 2^n$.

Every set A_ξ, member of Δ, meets, for some $\nu < cf(n) = \omega_\beta$ the set S_ν in a set $B_{\xi,\nu}$ of power $\geqslant p$ (since $p < m$). From $n^+ = 2^n$, and from $\omega_\beta = cf(n) \leqslant p < n < n^+$, it follows that for some $\nu' < cf(n) < n^+$, there are n^+ sets A_ξ, members of Δ, such that $|B_{\xi,\nu'}| \geqslant p$. For such (ξ,ν') let $C_{\xi,\nu'}$ be a subset of $B_{\xi,\nu'}$ of power p. So $C_{\xi,\nu'}$ is included in $A_\xi \cap S_{\nu'}$, and $C_{\xi,\nu'}$ belongs to $[S_{\nu'}]^p$. Therefore, from $|S_{\nu'}| < n$, and so $|[S_{\nu'}]^p| < n < n^+$ (this is a consequence of g.c.h.), it follows that there are two distinct sets $A_{\xi'}$ and $A_{\xi''}$, members of Δ, so that $C_{\xi',\nu'} = C_{\xi'',\nu'}$. So $|A_{\xi'} \cap A_{\xi''}| \geqslant p$, and we have a contradiction.

<u>Remark</u>. Under theorem hypotheses, if S is a set of power n, there is a subset L of $[S]^p$, of power $n^+ = 2^n$, such that for every distinct

members Z' and Z'' of L, we have $|Z' \cap Z''| < p$ (this result is in Tarski [4]).

7. Problem. We don't know if the theorem 2 is true whenever $p = \omega$, $m = \omega_1$, $n = \omega_2$ and $n^p = \omega_3 = 2^p$: we do not suppose g.c.h. .

REFERENCES

1. C. Berge, Graphes et hypergraphes, Dunod (1970) .
2. W. J. Frascella, Certain counterexamples to the construction of combinatorial designs on infinite sets, Notre-Dame journal of formal logic, 12, n°4, (1971), p. 461-466 .
3. K. Kuratowski and A. Mostowski, Set theory, North Holland and Polish scientific publishers (1968) .
4. A. Tarski, Sur la décomposition des ensembles en sous-ensembles presque disjoints, Fundamenta Mathematicae, 12, (1928), p. 188-205 ; 14, (1929), p. 205-215 .

INTERSECTING FAMILIES OF EDGES

IN HYPERGRAPHS HAVING THE HEREDITARY PROPERTY

V. Chvátal, Stanford University

I. Introduction. Let F be a hypergraph with vertex set

$$S = \{1, 2, \ldots, n\}$$

An intersecting family of edges in F is a partial hypergraph G such that

$$X, Y \in G \Rightarrow X \cap Y \neq \emptyset$$

If F is a simple graph, an intersecting family of edges is either a triangle or a star.

In a hypergraph F, the degree $\delta(x)$ of a vertex x is the number of edges containing x. Denote by

$$\delta(F) = \max_{x \in S} \delta(x)$$

the maximum degree in F. Clearly, the maximum size of an intersecting family is greater than or equal to $\delta(F)$.

Erdős, Chao-Ko and Rado have shown :

If F is complete r-uniform hypergraph with x vertices, $n \geq 2r$, then the maximum size of an intersecting family is equal to $\delta(F)$.

In this note, we use a similar technique to show that the same equality holds when the hypergraph F satisfies the following condition :

if $X_o \in F$, if $X \subseteq S$, and if there exists a one-to-one mapping f from X into X_o such that

$$f(x) \geq x \qquad (x \in X),$$

then $X \in F$.

2. Let X, Y be sets of positive integers. If there is a one-to-one mapping $f: X \to Y$ with $x \leq f(x)$ for each $x \in X$ then we write $X < Y$. A family G of sets will be called <u>intersecting</u> if $X \cap Y \neq \emptyset$ whenever $X,Y \in G$.

<u>Theorem</u>. <u>Let</u> F <u>be a family of subsets of</u> $\{1,2,\ldots,n\}$ <u>such that</u> $X \in F$, $Y < X \Rightarrow Y \in F$. <u>Let</u> G <u>be an arbitrary intersecting subfamily of</u> F. Then

$$|G| \leq |\{X \in F: 1 \in X\}| \quad . \tag{1}$$

<u>Proof</u>. We will proceed by induction on n; the case $n = 1$ is trivial. Now, let n be greater than one and let F, G satisfy the hypothesis of our theorem. To each family F^* of subsets of $\{1,2,\ldots,n\}$, we assign a weight $w(F^*) = \sum\sum k$ where the first sum runs over all $X \in F^*$ and the second one over all $k \in X$. Since we are going to prove (1), only the cardinality of G is of interest to us. Hence we may assume, without loss of generality, that G minimizes the weight among all the intersecting subfamilies of F having $|G|$ sets. First of all, we will prove that

$$X \in G, t \in X, s \notin X, s < t \Rightarrow (X - \{t\}) \cup \{s\} \in G \quad . \tag{2}$$

For this purpose, we will use the technique developed in [1]. Assume the contrary, i.e., let there be X, s, t violating (2). Fix s, t and set

$$G^* = \{Y \in G: t \in Y, s \notin Y, (Y - \{t\}) \cup \{s\} \notin G\} \quad .$$

Then $X \in G^*$. Moreover, let us set

$$H^* = \{(Y-\{t\}) \cup \{s\}: Y \in G^*\} ,$$

$$H = H^* \cup (G - G^*) .$$

Obviously, $|H| = |G|$, $H \subset F$ and $w(H) < w(G)$. By the minimality of $w(G)$, the family H cannot be intersecting. Since H^* and $G-G^*$ are both intersecting, there must be disjoint sets $Y \in H^*$ and $Z \in G-G^*$. Since $s \in Y$, we have $s \notin Z$. But $(Y-\{s\}) \cup \{t\} \in G^*$ and so $((Y-\{s\}) \cup \{t\}) \cap Z \neq \emptyset$. Therefore necessarily $t \in Z$. Since $Z \notin G^*$, we have $(Z-\{t\}) \cup \{s\} \in G$. Hence

$$\emptyset \neq ((Z-\{t\}) \cup \{s\}) \cap ((Y-\{s\}) \cup \{t\}) = (Y \cap Z) - \{s,t\}$$

contradicting $Y \cap Z = \emptyset$. Thus (2) is proved.

Next, let us note that, for any subsets X, Y of $\{1,2,\ldots,n\}$, $Y < X$ holds if and only if

$$|Y \cap \{k,k+1,\ldots,n\}| \leq |X \cap \{k,k+1,\ldots,n\}| \qquad (1 \leq k \leq n) .$$

Therefore

$$Y < X \Leftrightarrow \{1,2,\ldots,n\}-X < \{1,2,\ldots,n\}-Y . \tag{3}$$

Let us set

$$F_1 = \{X \in F: \{1,2,\ldots,n\}-X \in F\} .$$

From (3), we easily deduce that

$$X \in F-F_1 , Y < X \Rightarrow Y \in F-F_1 . \tag{4}$$

Indeed, $X \in F$ and $Y < X$ imply $Y \in F$. If $Y \notin F-F_1$ then necessarily $Y \in F_1$, i.e., $\{1,2,\ldots,n\}-Y \in F$. By (3), we then have $\{1,2,\ldots,n\}-X \in F$ contradicting $X \notin F_1$.

Now, set

$$F_2 = \{X \in F - F_1 : n \notin X\} \quad ,$$

$$F_3 = \{X \in F - F_1 : n \in X\} \quad ,$$

$$F_3^* = \{X - \{n\} : X \in F_3\} \quad .$$

From (4), it follows easily that

$$X \in F_2 , Y < X \Rightarrow Y \in F_2 \quad ,$$

$$X \in F_3^* , Y < X \Rightarrow Y \in F_3^* \quad .$$

We also set

$$G_i = G \cap F_i \qquad (i = 1,2,3) \quad ,$$

$$G_3^* = \{X - \{n\} : X \in G_3\} \quad .$$

and finally, let us set

$$H = \{X \in F : 1 \in X\} \quad ,$$

$$H_i = H \cap F_i \qquad (i = 1,2,3) \quad ,$$

$$H_3^* = \{X - \{n\} : X \in H_3\} \quad .$$

If $Y, Z \in G_3$ then $Y \cup Z \neq \{1,2,\ldots,n\}$ (otherwise $Y, Z \in F_1$). Therefore there is a $k \in \{1,2,\ldots,n-1\}$ with $k \notin Y$, $k \notin Z$. By (2), one has $(Y - \{n\}) \cup \{k\} \in G$ and so

$$(Y - \{n\}) \cap (Z - \{n\}) = ((Y - \{n\}) \cup \{k\}) \cup Z \neq \emptyset \quad .$$

Hence G_3^* is an intersecting subfamily of F_3^* .

Now, we can apply the induction step, obtaining thus

$$|G_2| \leq |H_2| \tag{5}$$

and

$$|G_3| = |G_3^*| \leq |H_3^*| = |H_3| \quad . \tag{6}$$

Finally, it is easy to see that

$$|G_1| \leq \tfrac{1}{2}|F_1| = |H_1| \quad . \tag{7}$$

Indeed, the family F_1 can be split into pairs $(X, \{1,2,\ldots,n\}-X)$. At most one element of each pair can be included in G_1 ; exactly one element of each pair is included in H_1 .

Summing up (5), (6) and (7) we obtain

$$|G| = |G_1| + |G_2| + |G_3| \leq |H_1| + |H_2| + |H_3| = |H|$$

which is the desired result. The proof is finished.

Perhaps the following strengthening of our theorem still remains valid:

Conjecture. Let F be a family of subsets of a finite set S such that $X \in F, Y \subset X \Rightarrow Y \in F$. Then there is a $t \in S$ such that every intersecting subfamily G of F satisfies

$$|G| \leq |\{X \in F:\ t \in X\}| \quad .$$

One might also believe that the following generalization of our theorem is true: Let F be a family of subsets of $\{1,2,\ldots,n\}$ such that $X \in F$, $Y < X \Rightarrow Y \in F$; let G be a subfamily of F containing no $k+1$ pairwise disjoint sets and such that $|G| > k$. Then

$$|G| \leq |\{X \in F: \{1,2,\ldots,k\} \cap X \neq \emptyset\}| \quad . \tag{8}$$

However, this statement is false whenever $k > 1$. Indeed, if F consists of all the subsets of $\{1,2,\ldots,2k+1\}$ then the right-hand side of (8) is $2^{2k+1}-2^{k+1}$. However, the family

$$G = \{X \subset \{1,2,\ldots,2k+1\}: |X| \geq 2\}$$

has no $k+1$ pairwise disjoint sets and includes

$$2^{2k+1} - (2k+2) > 2^{2k+1} - 2^{k+1}$$

sets. Nevertheless, it would be desirable to prove (8) under more restrictive conditions on F . Such a theorem might eventually imply the following number-theoretical conjecture of Erdös: Let S be a subset of $\{1,2,\ldots,m\}$ containing no $k+1$ pairwise coprime integers. Then $|S| \leq |T|$ where T is obtained by taking all those integers in $\{1,2,\ldots,m\}$ which are multiples of (at least one of) the first k primes.

REFERENCES

1. P. Erdös, Chao-Ko and R. Rado, "Intersection theorems for systems of finite sets," Quarterly J. of Math. (Oxford, 2nd sec.) 12 (1961), 313-320.

ON THEOREMS OF BERGE AND FOURNIER

W.H.Cunningham, University of Waterloo

Where A is a set and x, y are elements, $A + x$ denotes $A \cup \{x\}$ and $A - x$ denotes $A \setminus \{x\}$. In an expression such as $A - x + y$ we omit brackets, the "order of operations" being from the left. Two hypergraphs $H = (E_i | i \in I)$ and $H' = (F_i | i \in I)$ are <u>strongly isomorphic</u> if there exists a bijection f from the vertex set of H to the vertex set of H' such that $f(E_i) = F_i$ for each $i \in I$. That is, there exists an isomorphism which preserves the (given) correspondence of edges. Where n, h are positive integers and $n \geq h$, K_n^h denotes a simple hypergraph whose edges are precisely the h-element subsets of an n-element set. The following isomorphism theorem is the composite of results in [2] and [6].

<u>Theorem 1</u> (Berge and Fournier). <u>Let</u> $H = (E_i | i \in I)$ <u>be isomorphic to</u> K_n^h. <u>Let</u> $H' = (F_i | i \in I)$. <u>Suppose that</u>

(1) $$|E_i \cap E_j| = |F_i \cap F_j| \quad \text{for all } i,j \in I.$$

<u>Then</u> H' <u>is strongly isomorphic to</u> H <u>or to</u> $H'' = (X_0 \setminus E_i | i \in I)$ <u>where</u> $|X_0| = 2h$ <u>and</u> $X_0 \supseteq X$, <u>the vertex set of</u> H.

Using a result of [4] we will extend Theorem 1 to a considerably larger class of hypergraphs than the "complete" hypergraphs. A finite simple hypergraph $H = (E_i | i \in I)$ is here called a <u>matroid</u> if (2) is satisfied.

(2) If $i, j \in I$ and $x \in E_i \setminus E_j$, there exist $k \in I$ and $y \in E_j \setminus E_i$ such that $E_k = E_j + x - y$.

There are many hypergraphs associated with a matroid; we are choosing the one whose edges are the bases of the matroid. The reader may contrast this with [5], where the edges are the circuits of the matroid. Since every vertex of a hypergraph is an element of some edge, our matroids are "loopless". The rank function r of the hypergraph will coincide with the usual matroid rank function. Where X is the vertex set of the hypergraph H, $r(H)$ means $r(X)$. If there exists no non-empty proper subset A of X such that $r(A) + r(X \setminus A) = r(X)$, we say that H is <u>non-separable</u>. If H is isomorphic to K_n^h, then H is a matroid. Moreover, H is non-separable unless $h = 1$ or $h = n$, in which cases Theorem 1 is obvious. Thus Theorem 2 below is a generalization of Theorem 1.

<u>Theorem 2</u>. <u>Let</u> $H = (E_i | i \in I)$ <u>be a non-separable matroid and let</u> $H' = (F_i | i \in I)$. <u>Suppose that</u>

(3) $|E_i| = |F_i| = h$ <u>for each</u> $i \in I$, <u>and</u>

(4) $|E_i \cap E_j| = h - 1$ <u>if and only if</u> $|F_i \cap F_j| = h - 1$ <u>for all</u> $i, j \in I$.

<u>Then</u> H' <u>is strongly isomorphic to</u> H <u>or to</u> $H'' = (X_0 \setminus E_i | i \in I)$ <u>where</u> $|X_0| = 2h$ <u>and</u> $X_0 \supseteq X$, <u>the vertex set of</u> H.

We point out that (1) has been weakened to (3) and (4). Where $H = (E_i | i \in I)$ is a simple hypergraph, the line graph $L(H)$ of H, denotes the strict graph whose vertex set is $\{E_i | i \in I\}$ with E_i adjacent to E_j if and only if there exist $x \in E_i \setminus E_j$ and $y \in E_j \setminus E_i$ such that $E_j = E_i - x + y$. Thus (4) implies that $L(H)$ is isomorphic to $L(H')$. Necessary and sufficient conditions for two matroids to have isomorphic line graphs were obtained in [4]. A special case of this result is stated as Theorem 3.

The dual of the hypergraph $H = (E_i | i \in I)$ with vertex set X is $H^* = (X \setminus E_i | i \in I)$. This is not the usual hypergraph dual, but rather is motivated by matroid theory. It is well known that if H is a matroid, then so is H^*. An isthmus of $H = (E_i | i \in I)$ is a vertex which is an element of every E_i. We will say H' is "H plus isthmuses" if $H' = (A \cup E_i | i \in I)$ for some set A which is disjoint from the vertex set of H. The "strong isomorphism" in the statement of Theorem 3 is with respect to the correspondence of edges given by the graph isomorphism f of the hypothesis.

Theorem 3. *If H is a non-separable matroid and H' is a matroid and f is an isomorphism from $L(H)$ to $L(H')$, then H' is strongly isomorphic to H plus isthmuses or to H^* plus isthmuses.*

Now suppose that in Theorem 3 we also have $r(H') = r(H)$ and let X denote the vertex set of H. Then if H' is strongly isomorphic to $(A \cup E_i | i \in I)$ where $A \cap X = \phi$, we must have $A = \phi$. Otherwise H' will be strongly isomorphic to $(A \cup (X \setminus E_i) | i \in I)$ where $A \cap X = \phi$, whence $r(H) = |A| + |X| - r(H)$, so that $|A| = 2r(H) - |X|$. Thus to prove Theorem 2 from Theorem 3 we need only to prove that H' must be a matroid. This follows from Theorem 4 which characterizes line graphs of matroids among line graphs of hypergraphs.

Theorem 4. The finite simple hypergraph $H = (E_i | i \in I)$ is a matroid if and only if

(5) $L(H)$ is connected,

(6) Whenever E_1, E_2 are vertices at distance 2 in $L(H)$, there exist common neighbours E_3, E_4 of E_1, E_2 such that E_3 and E_4 are not adjacent.

Proof. Suppose that H is a matroid and that E_1, E_2 are vertices of $L(H)$ not joined by a path. Subject to this, choose $|E_1 \setminus E_2|$ to be minimum. Now $E_1 \neq E_2$, so there exists $x \in E_1 \setminus E_2$. Thus by (2) there exists $y \in E_2 \setminus E_1$ such that $E_3 = E_2 + x - y$ is a vertex of $L(H)$. Now $|E_1 \setminus E_3| < |E_1 \setminus E_2|$ so E_1 is joined to E_3 by a path in $L(H)$ and thus E_1 is joined to E_2. This is a contradiction, proving (5).

Next suppose E_1, E_2 are vertices of $L(H)$ at distance 2.

Then we may write $E_2 = E_1 - x + y - z + w$ where $E_1 - x + y$ is a vertex of $L(H)$ and $z \in E_1 - x$ and $w \notin E_1 + y$. Now if $E_1 - z + w$ is an edge of H, then $E_1 - x + y$, $E_1 - z + w$ are the required non-adjacent common neighbours of E_1 and E_2. Otherwise since $x \in E_1 \setminus E_2$, we must have $E_1 - z + y$ as an edge of H by (2). Similarly since $w \in E_2 \setminus E_1$, $E_1 - x + w$ must be an edge of H. Again we have the required common neighbours and (6) is proved.

Now suppose that H is finite and satisfies (5) and (6). Let k be a positive integer and consider the statement (7).

(7) If E_1, E_2 are at distance k in $L(H)$ and $x \in E_1 \setminus E_2$, there exists $y \in E_2 \setminus E_1$ such that $E_2 + x - y$ is a vertex of $L(H)$ at distance $k - 1$ from E_1.

When $k = 1$, (7) is obviously true. If $k = 2$, we may write $E_2 = E_1 - a + b - z + w$ where $E_1 - a + b$ is an edge of H and $z \in E_1 - a$, $w \notin E_1 + b$. Since the only possible common neighbours of E_1, E_2 are $E_1 - a + b$, $E_1 - a + w$, $E_1 - z + b$, and $E_1 - z + w$, then by (6) either $E_1 - z + w$ or both $E_1 - z + b$, $E_1 - a + w$ are edges of H. In either case we can check that (7) is satisfied, the only possible choices for x being a and z.

We now assume that (7) is satisfied for every $k \le n$ where n is an integer, $n \ge 2$. Suppose that E_1, E_2 are at distance $n + 1$ in $L(H)$ and that $x \in E_1 \setminus E_2$. We will show how to find a neighbour E_3 of E_2 at distance n from E_1 with $x \in E_3$. Let E_4 be a neighbour of E_2 at distance n from E_1. If $x \in E_4$, we set $E_3 = E_4$. Otherwise by hypothesis there exists $w \in E_4 \setminus E_1$ such that $E_5 = E_4 + x - w$ is a vertex of L(H) at distance n - 1 from E_1. Then since (7) is true for k = 2, there exists E_3 a common neighbour of E_5 and E_2, and thus at distance n from E_1, with $x \in E_3$. Now we may write $E_3 = E_2 + x - y$ for $y \in E_2$. If $y \in E_1$, then by hypothesis we can find $z \in E_3 \setminus E_1 = (E_2 \setminus E_1) + x - y$ such that $E_3 + y - z = E_2 + x - z$ is a vertex of L(H) at distance n - 1 from E_1. But then E_2 is at distance $\le n$ from E_1, a contradiction. Thus $y \in E_2 \setminus E_1$ and (7) is satisfied for k = n + 1. Since L(H) is finite and connected, this shows that H satisfies (2) and thus is a matroid. The proof is complete.

Remarks. A more general (but more complicated) result than Theorem 2 can be obtained by dropping the restriction to non-separable matroids and using the main result of [4] rather than Theorem 3. Notice that Theorem 4 does not require non-separability.

Berge [2] proves Theorem 1 when n is allowed to be an infinite cardinal number but h is not. This is done by proving the result for finite n and then invoking a result of [3]; namely, hypergraphs $H = (E_i | i \in I)$ and $H' = (F_i | i \in I)$ are strongly isomorphic if all corresponding pairs of finite partial hypergraphs are strongly isomorphic. This same approach will prove Theorem 2 for "finite-rank matroids". Of course the possibility that H' be strongly isomorphic to H'' in either Theorem 1 or Theorem 2 is excluded if H is infinite.

Further extensions of Theorem 1 would be interesting, but if we ask only that H be finite and $|E_i| = h$ for all $i \in I$, the problem will be difficult. In this case (1) is in general a much stronger statement than (3). It follows from (7) that (1) is equivalent (3) when H is a matroid. When this is not true the numbers $|E_i \cap E_j|$ for $i, j \in I$ seem to give far less information about H. The theory of combinatorial designs provides an interesting example. If the edges of H, H' are the blocks of symmetric (v, k, λ)-designs, then (1) is satisfied, since it is well-known that $|E_i \cap E_j| = \lambda$ if $i \neq j$. It is easy to check that (in non-trivial cases) h, H' are non-separable and that H' cannot be isomorphic to H^*. Yet for any integer $n \geq 3$, there exist at least $1 + 10[\frac{n}{5}]$ pairwise non-isomorphic symmetric $(2^n - 1, 2^{n-1} - 1, 2^{n-2} - 1)$-designs [7, page 409]. (As usual, $[x]$ denotes the largest integer less than or equal to x.)

Acknowledgment. I am grateful to Professor Jack Edmonds for several conversations on this subject, and especially for his suggestion that Theorem 2 could be proved.

REFERENCES

1. C. Berge, Graphes et hypergraphes, Dunod, Paris, 1970.

2. C. Berge, Une condition pour qu'un hypergraphe soit fortement isomorphe à un hypergraphe complet ou multiparti, C.R. Acad. Sc. Paris 274 (1972), 1783-1786.

3. C. Berge and R. Rado, Note on isomorphic hypergraphs and some extensions of Whitney's theorem to families of sets, to appear, J. Comb. Theory.

4. W. H. Cunningham, The matroids of a Basis Graph, not yet published.

5. W.H. Cunningham and J. Edmonds, this volume.

6. J.C. Fournier, Sur les isomorphismes d'hypergraphes, C.R. Acad. Sc. Paris 274 (1972), 1612-1614.

7. W.D. Wallis, Anne Penfold Street, Jennifer Seberry Wallis, Combinatorics : Room Squares, Sum-free Sets, Hadamard Matrices, Vol. 292, Lecture Notes in Mathematics, Springer-Verlag, Berlin-Heidelberg-New York, 1972.

EXTREMAL PROBLEMS ON GRAPHS AND HYPERGRAPHS

Paul Erdös, Hungarian Academy of Science

In this short survey I will state many solved and unsolved problems. I will give almost no proofs and will try to give extensive references, so that the interested reader can find what is omitted here. G_r denotes an r-graph, $G_r(k)$ an r-graph with k vertices and $G_r(k;m)$ an r-graph with k vertices and m r-tuples. $K_r(t)$ denotes the complete r-graph of t vertices, i.e. the r-graph $G_r(t;\binom{t}{r})$. $K_r(t,\dots,t)$ denotes the r-graph of rt vertices and t^r r-tuples where the vertices are split into r classes of t vertices each and every r-tuple contains one and only one vertex of each class.
If G_r is an r-graph then $f(n;G_r)$ is the smallest integer so that every $G_r(n;f(n;G_r))$ contains our G_r as a subgraph.
In 1940 Turán [1] proved that if $n \equiv s \pmod{t-1}$, then

$$f(n;K_2(t)) = \frac{t-2}{2(t-1)}(n^2 - s^2) + \binom{s}{2}.$$

He also proved that the only $G_2(n;K_2(t)-1)$ which does not contain a $K_2(t)$ is the complete $(t-1)$-partite graph $K_2(m_1,\dots,m_{t-1})$ where $m_1 + \dots + m_{t-1} = n$ and the summands are as nearly equal as possible. Turán's paper initiated the systematic study of extremal properties of graphs and hypergraphs. Turán posed the very beautiful and difficult problem of determining $f(n; K_r(t))$ for $r > 2$ and $t > r$. This problem is unsolved. It is not hard to see (Katona - Nemetz-Simonovits [2]) that

$$\lim_{n=\infty} f(n; K_r(t)) / \binom{n}{r} = c_{r,t}$$

always exists, but the value of $c_{r,t}$ is unknown for every $r > 2$, $t > r$ though Turán has some plausible conjectures.

In fact very few exact results are known for $r > 2$. Before I state systematically the problems and results in our subject I mention the following recent result of B. Bollobás who proved the following conjecture of Katona: Every $G_3(n; [\frac{n}{3}][\frac{n+1}{3}][\frac{n+1}{3}] + 1)$ contains three triples so that one of them contains the symmetric difference of the other two. The result is easily seen to be best possible. The paper of Bollobás will be published soon.

$f_r(n; k, l)$ is the smallest integer so that every $G_r(n; f_r(n; k, l)$ contains at least one $G_r(k; l)$ as a subgraph in other words the structure of our $G_r(k, l)$ is not specified. The study of $f_r(n; k, l)$ in general is simpler (but perhaps less interesting) than that of $f_r(n; G_r(k; l)$. In the first chapter I discuss $r = 2$ and in the second I state some of our meagre knowledge for $r > 2$

$r = 2.$

As far as I know the first paper which tried to study systematically extremal properties of graphs was [3]. First I state the following general theorem of Simonovits - Stone and myself [4]. Let G be a graph of chromatic number k. Then

(1) $$\lim_{n=\infty} f(n\,;G)/\binom{n}{2} = 1 - \frac{1}{k-1} .$$

In view of (1) we will mostly restrict ourselves to the study of bipartite graphs. A result of Kövari, the Turáns and myself states that [5] (the c's denote absolute constants not necessarily the same if they occur in different formulas)

(2) $$f(n\,;K_2(t,t)) < c_1 n^{2-\frac{1}{t}}$$

In other words every $G(n\,;[c_1 n^{2-\frac{1}{t}}])$ contains a complete bipartite graph $K_2(t,t)$ as a subgraph if c_1 is sufficiently large. We conjectured that (2) is best possible but this has been proved only for $t = 2$ and $t = 3$ [6]. Denote by C_k a circuit having k edges. Brown, V. T. Sós, Rényi and I proved that [6]

$$\lim_{n=\infty} f(n\,;C_4)/n^{3/2} = \frac{1}{2} .$$

Our proof in fact gives

(3) $$f(n\,;C_4) \le \frac{n^{3/2}}{2} + \frac{n}{4} + \sigma(n)$$

and in fact many of us conjectured that

$$f(n\,;C_4) = \frac{n^2}{4} + \frac{n}{4} + \sigma(n) .$$

Let $n = p^2 + p + 1$ where p is a power of a prime. Our method gives

(4) $$f(n\,;C_4) \ge \frac{(p+1)^2 p'}{2} + 1 .$$

It would be nice if we would have equality in (4).

I proved that

(5) $$f(n,C_{2k}) < c_1 n^{1+\frac{1}{k}}$$

I never published a proof of (5) since my proof was messy and perhaps even not quite accurate and I lacked the incentive to fix everything up since I never could settle various related sharper conjectures--all these have now been proved by Bondy and Simonovits--their paper will soon appear. Probably (5) is best possible but this has been proved only for $k = 2$ and $k = 3$ (Singleton). For further results on cycles see the papers of Bondy and Woodall [7].

Gallai and I proved that every $G(n ; [\frac{1}{2}(k - 1)n] + 1)$ contains a path of lenght k and V. T. Sós and I conjectured that every such graph contains every tree of k edges [1]. No progress has been made with this conjecture [8].

Let G be a bipartite graph. I conjectured that $f(n ; G)/n^{1+\alpha}$ tends to a finite non zero limit for some α of the form $\frac{1}{k}$ or $1 - \frac{1}{k}(k = 2,3,\ldots)$. Simonovits and I disproved this conjecture [9]. We still think that for every bipartite G there is an $\alpha, 1 < \alpha < 2$, for which

(6) $$\lim_{n=\infty} f(n ; G)/n^{\alpha} = c(G) \quad , \quad 0 < c(G) < \infty$$

but the set of these α's is everywhere dense in $(1,2)$. Probably the α in (6) is always rational.

Let G be the skeleton of a cube. Simonovits and I proved [9]

(7) $$f(n ; G) < cn^{8/5}$$

We could not decide whether (7) is best possible.

Simonovits and I determined $f(n\,;G)$ if G is the skeleton of an octahedron [10] and Simonovits determined $f(n\,;G)$ if G is the skeleton of the icosahedron

Before I close this chapter I state two simple unsolved questions considered by Simonovits and myself. Let G_k be the graph having the $1+k+\binom{k}{2}$ vertices $x_1\,;y_1,\ldots,y_k$, and $z_{i,j}$, $1 \le i < j \le k$. x_1 is joined to $y_1,\ldots,y_k$ and $z_{i,j}$ is joined to x_i and x_j. Is it true that

(8) $$f(n\,;G_k) < c_k n^{3/2}$$

I proved (8) for $k = 3$ [11]. G_k contains rectangles so that (8) if true is best possible. Denote by $G - x$ the graph obtained from G by removing the vertex x and all edges incident to it. Is it true that for every k

$$\lim_{n=\infty} f(n\,;G_k - x)/n^{3/2} = 0\,.$$

II

Now we discuss some problems and results for $r > 2$. A few years ago I proved that for every r and t there is an $\varepsilon_{r,t}$ so that every $G_r(n,[n^{r-\varepsilon_{r,t}}])$ contains a $K_r^{(r)}(t,\ldots,t)$ [12]. For $r = 2$ this is the theorem of Kövári and the Turáns stated in (2). For $r > 2$, $t \ge 2$ the exact value of $\varepsilon_{r,t}$ is not known. This result implies that every $G_r(n\,;[\varepsilon n^r])$ contains a subgraph of $m = m(n) \to \infty$ as vertices which has at least m^r/r^r edges. I conjecture that the following result is true: There is an absolute constant $c > \frac{1}{r^r}$ so that every $G_r(n;[\frac{n^r}{r^r}(1+\varepsilon)])$ contains a subgraph $G_r(m;[cm^r])$ where $m = m(n) \to \infty$ as $n \to \infty$. The case $r = 2$ is completely cleared up by the result of Stone and myself [13][4]. For $r > 2$ and for $r = 2$ and directed graphs or multigraphs many unsolved problems remain (see a forthcoming paper of Brown, Simonovits and myself).

In two forthcoming papers W. Brown, V. T. Sós and I began a systematic study of extremal problems for r-graphs. Before stating some of our results I state the most attractive unsolved problem:

Is it true that

$$f(n;G_3(6,3))/n^2 \to 0 \quad . \tag{9}$$

We proved $f(n;G_3(6;3)) > cn^{3/2}$ and it seems likely that in fact $f(n;G_3(6;3)) < n^{2-\varepsilon}$ for some $\varepsilon > 0$, but we could not even prove (9). Very recently Szemerédi states that he proved (9).

We prove that every $G_3(n;[c_1 n^{5/2}])$ contains a triangulation of the sphere for sufficiently large c_1 (the result fails if c_1 is small). Simonovits independently proved that every $G_3(n;[cn^{3-\frac{1}{k}}])$ contains a k- tuple pyramid and that for $k = 2$ and $k = 3$ the exponent is best possible.

To conclude I state some of the problems, results, for 3-graphs. In our paper for simplicity we take $r = 3$ (some of the results hold with appropriate change for $r > 3$). We have

$$\lim_{n=\infty} \frac{1}{n^2} f(n;G_3(4,2)) = \frac{1}{6} \quad .$$

but the determination of

$$\lim_{n=\infty} \frac{1}{n^3} f(n;G_3(4;3))$$

seems to be very difficult, perhaps as difficult as Turán's problem on $f(n;K_3(4))$.

(10) $$c_1 n^{5/2} < f(n;G_3(5;4)) < c_2 n^{5/2}$$

We have not been able to get an asymptotic formula for $f(n;G_3(5;4))$.

By the probabilistic method we proved

(11) $$f(n;G_3(k,k-1)) > n^{2+\varepsilon_k}$$

but except for $k = 5$ we do not know the exact value of ε_k . It is easy to see that for every $k > 3$ 0 ,

(12) $$c_1^{(k)} n^2 < f(n;G_3(k,k-2)) < c_2^{(k)} n^2 .$$

As stated previously $c_1^{(k)} = c_2^{(k)} = \frac{1}{6}$, but for $k > 4$ we have no asymptotic formula for $f(n;G_3(k,k-2))$. I would not be surprised if it would turn out that for every k

(13) $$\lim_{n=\infty} \frac{1}{n^2} f(n;G_3(k,k-2)) = \frac{1}{6} .$$

The only argument in favor of this conjecture (the conjecture may easily turn out to be nonsense) is the following

Theorem. Every $G_3(n;[\frac{1}{3}\binom{n}{2}]+1)$ contains either a $G_3(5;3)$ or a $G_3(6;4)$.

Let $x_1,\dots,x_n$ be the vertices of our graph and $T_1,\dots,T_\ell$, $\ell = [\frac{1}{3}\binom{n}{2}]+1$ its triples. Since $3\ell > \binom{n}{2}$ at least one pair say

(x_1, x_2) is contained in two triples say T_1 and T_2. We can clearly assume that no pair is contained in three triples for otherwise our graph would contain a $G_3(5;3)$. Also if say $T_1 = (x_1, x_2, x_3)$ and $T_2 = (x_1, x_2, x_4)$, no T_i can contain (x_3, x_4) since if $T_i = (x_3, x_4, x_5)$, then $G_3(x_1, x_2, x_3, x_4, x_5)$ contains three triples. Thus to every pair which is contained in two triples there corresponds a pair not contained in any triple. This correspondence can not be one to one since otherwise ℓ would be x at most $[\frac{1}{3}\binom{n}{2}]$. Thus there must be two triples (x_1, x_5, x_6) and (x_2, x_5, x_6) which have a common pair and also exclude (x_1, x_2). But then $G_3(x_1, \ldots, x_6)$ contains four triples and thus our theorem is proved. I hope this argument can be improved.

It is clear that many more problems could be formulated.

REFERENCES

1. P. Turán, Eine extremalaufgabe aus der Graphentheorie (in Hungarian). Mat és Fiz Lapok 48(1941), 436-452 see also Colloquium Math. 3(1954), 19-30.

2. G. Katona, T. Nemetz and M. Simonovits, On a graph problem of P. Turán, Mat. Lapok 15(1964), 228-238 (in Hungarian). See also Y. Spencer, Turáns theorem for k-graphs, Discrete Math. 2(1972), 183-186.

3. P. Erdös, Extremal problems in graph theory, Theory of graphs and its applications, Proc. Symp. held at Smolenice, June, 1963, Publishing House of Czechoslovak Academy and Academic Press, 29-36.

4. P. Erdös and A. H. Stone, On the structure of linear graphs, Bull. Amer. Math. Soc. 52(1946), 1087-1091, P. Erdös and M. Simonovits, A limit theorem in graph theory, Studia Sci. Mat. Hung. Acad. 1(1966), 51-57.

5. T. Kövári, V. T. Sós and P. Turán, On a problem of K. Zarankiewicz, Colloquium Math. 3(1954), 50-57.

6. W. G. Brown, On graphs that do not contain a Thomsen graph, Canad. Math. Bull 9(1966), 281-285, P. Erdös, A. Rényi and V. T. Sós, On a problem of graph theory, Studia Sci. Math. Hung. Acad. 1(1966), 215-235.

7. J. A. Bondy, Large cycles in graphs, Discrete Math. 1(1971), 121-132, also Panyclic graphs I, J. Combinatorial Theory 11(1971), 80-84, II will appear soon. D. R. Woodall, Sufficient conditons for circuits in graphs, Proc. London Math. Soc 24(1972), 739-755.

8. P. Erdös and T. Gallai, On the maximal paths and circuits of graphs, Acta. Math. Hung. Acad. Sci. 10(1959), 337-356, for some sharper results see B. Andrásfai, On the paths circuits and loops of graphs, Mat. Lapok 13(1962) (in Hungarian).

9. P. Erdös and M. Simonovits, Some extremal problems in graph theory, Combinatorial Theory and its applications, Colloquium held in Balatonfüred, Hungary 1969, North Holland Publishing Company 1970, 377-390.

10. P. Erdös and M. Simonovits, An extremal graph problem, Acta. Math. Acad. Sci. Hungar. 22(1971), 275-282.

11. P. Erdös, On some extremal problems in graph theory, Israel J. Math. 3(1965), 113-116.

For further papers on extremal problems on graphs see P. Erdös, On some near inequalities concerning extremal properties of graphs, and M. Simonovits, A method for solving extremal problems in graph theory. Stability problems. Theory of Graphs Proc. Coll. Tihany, Hungary, 1966, Acad. Press 77-81 and 279-319.

12. P. Erdös, On extremal problems of graphs and generalized graphs, Israel J. Math. 2(1964), 183-190, see also ibid 251-261.

13. P. Erdös, On some extremal properties of r-graphs, Discrete Math. 1(1971), 1-6.

14. W. G. Brown, P. Erdös and V. T. Sós, On the existence of triangulated spheres in 3-graphs, and related problems, will appear soon .

HYPERGRAPH RECONSTRUCTION

Vance Faber, University of Colorado

1. For the general theory of hypergraphs see [1]. If $(E_i: i \in M)$, denotes a non-empty family of non-empty sets E_i indexed by M then, for every $I \subseteq M$, we put

$$E_I = \bigcup (i \in I) E_i.$$

If $I \neq \emptyset$, we put

$$E_{[I]} = \bigcap (i \in I) E_i.$$

A similar notation is used when E is replaced by another letter. A <u>hypergraph</u> is a pair

$$H = (X, (E_i: i \in M)),$$

where $X = E_M = \{x_i: i \in N\}$. The elements of X are called the <u>vertices</u> of H, and the sets E_i are the <u>edges</u> of H. The <u>order</u> of H is $n(H) = |H|$, and the <u>rank</u> of H is $\sup\{|E_i|: i \in M\}$ where, for every set S, we denote by $|S|$ the cardinality of S. Let $m(H) = |M|$. If $|E_i| = r$ for all $i \in M$, the hypergraph H is said to be <u>r-uniform</u>. The <u>valency</u> of a vertex $x \in E_M$ is $d(x) = |\{E_i: i \in M, x \in E_i\}|$. We let $\delta(H) = \sup\{d(x): x \in E_M\}$. If $d(x) = \delta$ for all $x \in E_M$, the hypergraph H is said to be <u>δ-regular</u>.

Two hypergraphs $H = (E_i: i \in M)$ and $H' = (F_i: i \in M)$ are said to be <u>isomorphic</u> if there exists a bijection $\varphi: E_M \to F_M$ and a permutation π of M such that

$$\varphi(E_i) = F_{\pi(i)} \quad (i \in M).$$

If H and H' are isomorphic, we write

$$H \simeq H'.$$

The hypergraphs H and H' are called <u>strongly isomorphic</u> if there exists a bijection $\varphi: E_M \to F_M$ such that

$$\varphi(E_i) = F_i \ (i \in M).$$

Strong isomorphism is expressed by

$$H \cong H'.$$

The hypergraphs H and H' are called <u>equivalent</u> if $E_M = \{x_i: i \in N\}$ and $F_M = \{y_i: i \in N\}$ and there exists a permutation π of M such that the mapping $\varphi(x_i) = y_i$ has the property

$$\varphi(E_i) = F_i \ (i \in M).$$

If H and H' are equivalent, we write

$$H \equiv H'.$$

Let $(E_i: i \in M)$ be a hypergraph. We shall discuss several types of subhypergraphs. Suppose $E_M = \{x_i: i \in N\}$. If $A \subset N$, the <u>hypergraph generated by A</u> is

(1) $$H|A = (E_i \cap \{x_i: i \in A\} \neq \phi: \ i \in M).$$

The <u>section hypergraph generated by A</u> is

(2) $$H \times A = (E_i: i \in M, E_i \subseteq \{x_i: i \in A\}).$$

If $A \subset M$, the <u>partial hypergraph generated by A</u> is

(3) $$H\,(E_i: i \in A) = (E_i: i \in A).$$

The <u>section partial hypergraph generated by A</u> is the hypergraph

(4) $$H|(E_M - E_{M-A}).$$

It is known that the types (1) and (3) are dual, as are the types (2) and (4).

If $A \subseteq B$, then it is clear that $(H|B)|A = H|A$. It is also easy to show that a similar equation holds for the types (2), (3) and (4). To represent all four of these equations by one equation we adopt the notation $H|i|A$ (i = 1, 2, 3, 4) to represent the hypergraph of type (i) (i = 1, 2, 3, 4). For example:

<u>Lemma 1</u>. If $A \subseteq B$, $(H|i|B)|i|A = H|i|A$.

<u>Proof</u>. Only the type (4) is not completely trivial, where it is necessary to verify that if $A = B - S$, then

$$E_M - E_{M-(B-S)} = E'_M - E'_{M-(B-S)}$$

where $E'_j = E_j \cap (E_M - E_{M-B})$ for all $j \in M$.

2. In this paper we shall discuss several generalizations of Ulam's reconstruction conjecture [9; p. 29]. (For a general survey of this conjecture see [5] and [8].)

Let H and H' be hypergraphs of rank at most r. Corresponding to the hypergraphs of type (1) and (2), H and H' are said to satisfy the <u>$U^n_{k,r}(i)$ hypothesis</u> (i = 1, 2) if H and H' have order n and $H|i|N - A \simeq H'|i|N - A$ for every k-element subset A of N. Corresponding to the hypergraphs of type (3) and (4), H and H' are said to satisfy the <u>$U^m_{k,r}(i)$ hypothesis</u> (i = 3,4) if H and H' have m edges and $H|i|(M - A) \simeq H'|i|(M - A)$ for every k-element subset A of M.

<u>The $U^n_{k,r}(i)$ conjectures (i = 1,2)</u>: If the hypergraphs H and H' satisfy the $U^n_{k,r}(i)$ hypothesis (i = 1, 2) with $n \geq k + r$, then $H \simeq H'$.

<u>The $U^m_{k,r}(3)$ conjecture</u>: If the hypergraphs H and H' satisfy the $U^m_{k,r}(3)$ hypothesis with $2^{m-1-k} > r$, then $H \simeq H'$. (No simple conditions seem to be available for a $U^m_{k,r}(4)$ conjecture.)

Each of these conjectures has a logically equivalent dual. We denote by $V^m_{k,\delta}(i)$ (i = 3, 4) the dual to $U^n_{k,r}(i)$ (i = 1, 2) and by $V^n_{k,\delta}(1)$ the dual to $U^m_{k,r}(3)$. (Recall that (1) and (2) are the duals to (3) and (4), respectively.)

In the next section, we shall show that the $U^n_{k,r}(i)$ conjectures (i = 1, 2) hold if $k \geq r$. (actually a much stronger theorem will be proven.) This will generalize a theorem of P. Kelly [7] for the case $r = 2$.

Some of the results of this paper were announced in [3].

3. The first theorem we shall prove is a generalization of what is usually called Kelly's Lemma [5], [6]. Let

$$\alpha_i(S, K, H) = |\{H|i|A\colon A \supseteq S,\ H|i|A \simeq K\}|.$$

Loosely speaking, this is the number of hypergraphs of the type (i) in H containing S and isomorphic to K. It is convenient to let $p = n$ and $P = N$ when $i = 1$ or 2, and to let $p = m$ and $P = M$ when $i = 3$ or 4. If $A \subseteq P$, we let $\overline{A} = P - A$. Also, let $\alpha_i(K, H) = \alpha_i(\phi, K, H)$.

<u>Theorem 1</u>. If H and H' satisfy the $U^p_{k,r}(i)$ hypothesis and if K is a hypergraph with $p(K) = t \le p - k$, then

$$\alpha_i(S, K, H) = \alpha_i(S, K, H') \quad (i = 1, 2, 3, 4)$$

for all $S \subseteq P$ with $|S| \le k$.

<u>Proof</u>. We shall suppress the index i throughout this proof. By the principle of inclusion-exclusion, we have

$$\alpha(K, H\|\overline{A}) = \sum_{S \subseteq A} (-1)^{|S|}\alpha(S, K, H).$$

Note that

(a)
$$\sum_{\substack{|A| = k \\ A \subseteq P}} \alpha(K, H\|\overline{A}) = \binom{p - t}{k} \alpha(K, H).$$

Thus if $|S| = s$,

(b)
$$\sum_{\substack{A \supseteq S \\ |A| = k}} \alpha(K, H\|\overline{A}) = \sum_{\substack{A \supseteq S \\ |A| = k}} \alpha(K, (H\|\overline{S})\|\overline{(A - S)})$$

$$= \binom{p - t - s}{k - s} \alpha(K, H\|\overline{S})$$

$$= \binom{p - t - s}{k - s} \sum_{T \subseteq S} (-1)^{|T|}\alpha(T, K, H).$$

By the $U^p_{k,r}$ hypothesis,

(c) $$\alpha(K,\ H\|\bar{A}) = \alpha(K,\ H'\|\bar{A})$$

for all $A \subset P$ with $|A| = k$. Since $p - t \geq k$, (c) implies that $\alpha(K, H) = \alpha(K, H')$. Inductively, assume that $\alpha(R, K, H) = \alpha(R, K, H')$ for all $R \subseteq P$ with $|R| < s \leq k$. Then since $p - t - s \geq k - s \geq 0$, (b) yields

$$\alpha(S,\ K,\ H) = \alpha(S,\ K,\ H')$$

for all $S \subseteq P$ with $|S| = s$.

Remark 1. Note that the full strength of the $U^p_{k,r}$ hypothesis was not used in this theorem. All that was needed was the equation (c).

Corollary 1. Suppose H and H' are r-uniform hypergraphs of order $n \geq 2r$. Let k be an integer such that $r \leq k \leq n - r$.

(1) $m(H|\bar{A}) = m(H'|\bar{A})$ for all $A \subseteq N$ with $|A| = k$ implies that $H \equiv H'$.

(2) $m(H \times \bar{A}) = m(H' \times \bar{A})$ for all $A \subseteq N$ with $|A| = k$ implies that $H \equiv H'$.

In particular, the $U^n_{k,r}(i)$ conjectures $(i = 1, 2)$ hold when $k \geq r$.

Proof. Let K be a single r-tuple. Then $\alpha_2(K, H \times \bar{A}) = m(H \times \bar{A})$, so $\alpha_2(S, K, H) = \alpha_2(S, K, H')$ for all S with $|S| = r$ means $\{x_i : i \in S\}$ is an edge of H if and only if $\{y_i : i \in S\}$ is an edge of H'. Thus statement (2) follows from Theorem 1 (see Remark 1.) Since

(*) $m(H|\bar{A}) = m(H) - m(H \times A)$, we have

$$\sum_{|A| = k} m(H|\bar{A}) = \sum_{|A| = k} m(H) - \sum_{|A| = k} m(H \times A)$$

$$= \left[\binom{n}{k} - \binom{n-r}{k-r}\right] m(H).$$

It follows that statement (1) implies that $m(H) = m(H')$ and hence from (*) that statement (2) holds.

Corollary 2. Suppose H and H' are δ-regular hypergraphs with $m \geq 2\delta$ edges. Let k be an integer such that $\delta \leq k \leq m - \delta$.

(3) $|E_{\bar{A}}| = |F_{\bar{A}}|$ for all $A \subseteq M$ with $|A| = k$ implies that $H \cong H'$.

(4) $|E_M - E_A| = |F_M - F_A|$ for all $A \subseteq M$ with $|A| = k$ implies that $H \cong H'$.

In particular, the $V^m_{k,\delta}(i)$ conjectures $(i = 3, 4)$ hold when $k \geq \delta$.

Proof. This is the dual of Corollary 1.

4. The cases $k < r$. We have only limited knowledge about the cases $k < r$ (except $k = 1$.) In Lemma 2, we prove a theorem which was proved for graphs [5] by a different method.

Lemma 2. If H and H' are hypergraphs of rank r with $2^{m-2} > r$ and if $(E_i: i \in M, i \neq j) \simeq (F_i: i \in M, i \neq j)$ for all $j \in M$, then $|E_M| = |F_M|$ and

$$|\{x \in E_M: d(x) = t\}| = |\{y \in F_M: d(y) = t\}|$$

for all $t > 0$. That is, the $U^m_{1,r}(3)$ hypothesis with $2^{m-2} > r$ implies that H and H' have the same set of valencies.

Proof. Suppose $|E_M| = |F_M| + s$ with $s \geq 0$. Since $|E_{\overline{\{j\}}}| = |F_{\overline{\{j\}}}|$,

$$|\{x \in E_j: d(x) = 1\}| = |\{y \in F_j: d(y) = 1\}| + s$$

and

$$|\{x \in E_M: d(x) = 1\}| = |\{y \in F_M: d(y) = 1\}| + ms.$$

Since

$$|\{x \in (E_i: i \neq j): d(x) = t\}| = |\{x \in E_M: d(x) = t\}| - |\{x \in E_j: d(x) = t\}| + |\{x \in E_j: d(x) = t + 1\}|,$$

by induction,

$$|\{x \in E_j: d(x) = t\}| + (-1)^t s\binom{m}{t}\frac{t}{m} = |\{y \in F_j: d(y) = t\}|$$

and

$$|\{x \in E_M: d(x) = t\}| + (-1)^t s \binom{m}{t} = |\{y \in F_M: d(y) = t\}|.$$

Then

$$mr \geq \sum\{d(x_i): x_i \in E_M\} \geq s[\binom{m}{1} + 3\binom{m}{3} + \ldots]$$

$$= sm[\binom{m-1}{0} + \binom{m-1}{2} + \ldots] = sm2^{m-2}.$$

Thus $r \geq s\, 2^{m-2}$. Since $2^{m-2} > r$, $s = 0$.

Remark 2. It is clear by induction that the $U^m_{k,r}(3)$ hypothesis with $2^{m-1-k} > r$ implies that H and H' have the same set of valencies. In fact, the same method can be used to show that the $U^p_{1,r}(i)$ conjecture ($i = 1, 2, 3$) is stronger than the $U^p_{k,r}(i)$ conjecture ($i = 1, 2, 3$) for any $k > 1$, as we now show.

Theorem 2. The $U^p_{1,r}(i)$ conjecture ($i = 1, 2, 3$) implies the $U^p_{k,r}(i)$ conjecture ($i = 1, 2, 3$) for all k.

Proof. Let k be an integer greater than 1. Suppose that the $U^p_{t,r}$ conjecture (we suppress the index i) holds for all $t < k$. Suppose H and H' satisfy the $U^p_{k,r}$ hypothesis. Let $j \in P$, then for each $k - 1$ element subset $A - \{j\}$ of $\overline{\{j\}}$,

$$(H\|\overline{\{j\}})\|\overline{A - \{j\}} = H\|\overline{A} \simeq H'\|\overline{A} = (H'\|\overline{\{j\}})\|\overline{A - \{j\}}.$$

Thus by the $U^{p-1}_{k-1,r}$ conjecture, $H\|\overline{\{j\}} \simeq H'\|\overline{\{j\}}$. Hence the $U^p_{1,r}$ conjecture implies that $H \simeq H'$.

Remark 3. Many of the theorems concerning the reconstruction of graphs which follow from Kelly's Lemma can be proved in identical fashion for hypergraphs from Theorem 1. We mention only two of these. We wish to provide evidence for the conjecture that the $U^p_{1,r}(i)$ conjectures ($i = 1, 2, 3$) are logically equivalent. It is clear that the $U^n_{1,r}(1)$ conjecture and the $U^m_{1,r}(3)$ conjecture are intimately related since the dual of $U^n_{1,r}(1)$, namely $V^m_{1,\delta}(3)$, has conditions quite similar to $U^m_{1,r}(3)$. Theorems 3 and 4 relate the $U^n_{1,r}(2)$ and $U^m_{1,r}(3)$ conjectures.

<u>Theorem 3</u>. If the hypergraphs H and H' satisfy the $U^m_{1,r}(3)$ hypothesis with $2^{m-2} > r$, then they satisfy the $U^n_{1,r}(2)$ hypothesis.

<u>Proof</u>. By Lemma 2, H and H' have the same set of valencies. It is now a simple matter to use Theorem 1 to show that H and H' satisfy the $U^n_{1,r}(2)$ hypothesis. For details, see [4] where this theorem is proved for graphs.

It is known that the hypergraphs $H = (E_i : i \in M)$ and $H' = (F_i : i \in M)$ are strongly isomorphic if and only if $|E_{[I]}| = |F_{[I]}|$ for all nonempty $I \subseteq M$. The <u>k-representing hypergraph</u> $R_k(H)$ of a hypergraph $H = (E_i : i \in M)$ is, by definition, the hypergraph with vertex set M in which a set $I \subseteq M$ forms an edge of multiplicity t if and only if $|E_{[I]}| = t$ and $1 \leq |I| \leq k$.

Associated with each finite nonempty set M, Berge and Rado [2] have defined the hypergraphs K(M) and L(M) as follows: K(M) is the dual of the hypergraph formed by the set of all nonempty subsets of M with the same parity as M; L(M) is the dual of the hypergraph formed bythe set of all nonempty subsets of M with parity opposite to that of M. We need the following lemma, which follows immediately from [2, Theorem 3].

<u>Lemma 3</u>. Let k be an integer $2 \leq k < m$. Let $H = (E_i : i \in M)$ and $H' = (F_i : i \in M)$ be hypergraphs with $|M| = m$. Suppose that there are no sets A, B, I with $A \subseteq E_M$, $B \subseteq F_M$, and $I \subseteq M$ such that $|I| = k + 1$ and the two hypergraphs $(A \cap E_i : i \in I)$, $(B \cap F_i : i \in I)$, in any order, are isomorphic to K(I), L(I) respectively. If $R_k(H) \simeq R_k(H')$, then $H \simeq H'$.

__Theorem 4__. Let H, H', k and m be as in Lemma 3. The following two statements are equivalent:

(a) If H and H' satisfy the $U^m_{1,r}(3)$ hypothesis, then they are isomorphic.

(b) If $R_k(H)$ and $R_k(H')$ satisfy the $U^m_{1,r}(2)$ hypothesis, then they are isomorphic.

__Proof__. It is easy to see that for any hypergraph $H = (E_i : i \in M)$, $R_k(E_i : i \neq j) = R_k(H) \times \{\bar{j}\}$. If $(E_i : i \neq j) \simeq (F_i : i \neq j)$ for all $j \in M$, we have

$$R_k(H) \times \{\bar{j}\} = R_k(E_i : i \neq j) \simeq R_k(F_i : i \neq j) = R_k(H') \times \{\bar{j}\}$$

for all $j \in M$. Thus $R_k(H) \simeq R_k(H')$ and by Lemma 3,

$$H \simeq H'.$$

If $R_k(H) \times \{\bar{j}\} \simeq R_k(H') \times \{\bar{j}\}$ for all $j \in M$, we have $R_k(E_i : i \neq j) \simeq R_k(F_i : i \neq j)$ for all $j \in M$. Thus by Lemma 3, $(E_i : i \neq j) \simeq (F_i : i \neq j)$ for all $j \in M$, so $H \simeq H'$. Thus

$$R_k(H) \simeq R_k(H').$$

__Remark 4__. It is possible that some modification of the proof of Theorem 4 could be used to show that the $U^m_{1,r}(3)$ conjecture implies the $U^n_{1,r}(2)$ conjecture, but the author has been unable to find it.

REFERENCES

1. C. Berge, Graphes et Hypergraphes, Dunod, Paris, 1970.

2. C. Berge and R. Rado, "Note on isomorphic hypergraphs and some extensions of Whitney's Theorem to families of sets," to appear in Journal Of Combinatorial Theory.

3. V. Faber, "Reconstruction of graphs from indexed p - 2 point subgraphs," Notices of Amer. Math. Soc. 18(1971)807.

4. D. L. Greenwell, "Reconstructing graphs," Proc. Amer. Math. Soc. 30(1971)431-433.

5. D. L. Greenwell and R.L. Hemminger,"Reconstructing graphs," The Many Facets of Graph Theory (G. T. Chartrand and S. F. Kapoor, eds.) Springer-Verlag, New York, 1969.

6. P. J. Kelly, "A congruence theorem for trees," Pac. J. Math. 7(1957)961-968.

7. _________, "On some mappings related to graphs," Pac. J. Math. 14(1964)191-194.

8. P. V. O'Neil, "Ulam's conjecture and graph reconstructions," Am. Math. Monthly 77(1970)35-43.

9. S. M. Ulam, A Collection of Mathematical Problems, Wiley (Interscience), New York, 1960.

UNE CONDITION POUR QU'UN HYPERGRAPHE, OU SON COMPLEMENTAIRE, SOIT FORTEMENT ISOMORPHE A UN HYPERGRAPHE COMPLET

J.C. Fournier, Université Paris VI

On désigne par K_n^h l'hypergraphe simple dont les arêtes sont les h-parties d'un ensemble à n éléments. Pour toutes notions et notations sur les isomorphismes d'hypergraphes on renvoie à [1]. Le théorème suivant est dû à C. Berge [1] pour le cas $n > 2h$ et Fournier [3] pour le cas $n \leq 2h$. La démonstration, par récurrence sur h, que nous en donnons ici rassemble ces 2 cas*.

<u>Théorème</u>. <u>Soient deux hypergraphes</u>

$$H = (E_i / i \in M) \simeq K_n^h$$

<u>et</u> $H' = (F_i / i \in M)$ <u>h-uniforme, tels que</u>

$$(1) \qquad |F_i \cap F_j| = |E_i \cap E_j| \qquad (i,j \in M)$$

<u>Si</u> $H' \not\cong H$, <u>alors</u>

$$H' \cong (X_o - E_i / i \in M) ,$$

<u>où</u>

$$X_o \supset X \ \text{ et } \ |X_o| = 2h \qquad (X = \bigcup_{i \in M} E_i) .$$

<u>Lemme</u>. <u>Soient deux hypergraphes, $H = (E_i / i \in M)$ h-uniforme, et $H' = (F_i / i \in M)$ h'-uniforme, tels que</u>

$$\cdot \ |F_i \cap F_j| = h'-1 \Longleftrightarrow |E_i \cap E_j| = h-1 \qquad (i,j \in M)$$

<u>Soient 3 arêtes E_i, E_j et E_k s'intersectant deux à deux en h-1 sommets. On a</u>

<u>- soit $F_k \supset F_i \cap F_j$, et alors pour tout $\ell \in M$ tel que E_ℓ intersecte E_i, E_j et E_k en h-1 sommets on a</u> $F_\ell \supset F_i \cap F_j$ /

(*) Ce théorème a été généralisé par W.H.Cunningham, cf. [2]. On peut noter que notre démonstration se généralise au théorème 3 [2] de Cunningham qui l'a conduit à cette généralisation.

- <u>soit $F_k \subset F_i \cup F_j$, et alors pour tout ℓ comme précédemment on a</u> $F_\ell \subset F_i \cup F_j$.

<u>Démonstration du lemme.</u> F_i, F_j et F_k s'intersectent 2 à 2 en $h'-1$ sommets. Posons $B = F_i \cap F_j$ et $B' = F_i \cup F_j$. Si $F_k \not\supset B$, soit $b \in B-F_k$. Comme $|F_k \cap F_i| = h'-1$ et $B \subset F_i$ nécessairement $F_k \supset F_i-b$. De même $F_k \supset F_j-b$. Or $|(F_i-b) \cup (F_j-b)| = |B'-b| = h'$, par conséquent $F_k = B'-b \subset B'$. Supposons maintenant $F_k \supset B$ et $F_\ell \not\supset B$; comme pour F_k précédemment on a, si $b \in B-F$, $F_\ell = B'-b$; mais alors $|F_k \cap F_\ell| = |B-b| < h'-1$, contradiction qui montre que nécessairement $F_\ell \supset B$. De même lorsque $F_k \subset B'$ on montre que $F_\ell \subset B'$.

<u>Démonstration du théorème.</u> Posons $S = \bigcap_{i \in M} F_i$, $G_i = F_i - S$ $(i \in M)$, $H'' = (G_i / i \in M)$ et $\overline{E}_i = X-E_i$, $\overline{H} = (\overline{E}_i / i \in M)$. Il suffit de montrer que $H'' \simeq H$ ou $\overline{H}$.

H'' est h''-uniforme $(h'' = h-|S|)$ et vérifie

$$(2) \qquad \bigcap_{i \in M} G_i = \emptyset$$

$$(3) \qquad |G_i \Delta G_j| = |E_i \Delta E_j| \qquad (i,j \in M)$$

(Δ désigne la différence symétrique).

Si $n \leq 2h$ on peut remplacer H par $\overline{H}$ sans que cela ne change (3) ni la conclusion $H'' \cong H$ ou $\overline{H}$, et l'on aura $n \geq 2\overline{h} = 2(n-h)$. On peut donc toujours supposer $n \geq 2h$. Montrons par récurrence sur h de 1 à $[\frac{n}{2}]$ que si $H \simeq K_n^h$ et H'' est h''-uniforme et vérifie (2) et (3) alors $H'' \simeq H$ ou $\overline{H}$.

Par application du lemme à H et H'' (ce qui est possible grâce à (3)) le cas $h=1$ se traite sans difficulté. Supposons donc $h > 1$ et la proposition vraie pour $h-1$.

1). Soit $A_o \subset X$, $|A_o| = h-1$. On a alors l'un des 2 cas suivants :

<u>1er cas</u> : il existe $B_o \subset Y$, $|B_o| = h''-1$ tel que

$$E_i \supset A_o \Longleftrightarrow G_i \supset B_o \quad (i \in M)$$

<u>2ème cas</u> : il existe $B'_o \subset Y$, $|B'_o| = h''-1$ tel que

$$E_i \supset A_o \Longleftrightarrow G_i \subset B'_o \quad (i \in M)$$

$$(X = \bigcup_{i \in M} E_i \quad \text{et} \quad Y = \bigcup_{i \in M} G_i).$$

En effet comme $n \geq 2h$ et $h \geq 2$ il existe au moins 3 arêtes de H contenant A_o. D'après le lemme il existe : soit B_o tel que $E_i \supset A_o \Longrightarrow G_i \supset B_o$, soit B'_o tel que $E_i \supset A_o \Longrightarrow G_i \subset B'_o$. Dans ces 2 cas on a également les implications inverses par application du lemme en échangeant H et H'' et en observant que les arêtes $G_i \supset B_o$ (ou $G_i \subset B'_o$) s'intersectent 2 à 2 en $h''-1$ sommets.

Si l'on a le 2ème cas, il suffit de remplacer H'' par $\overline{H}''$, ce qui ne change pas les hypothèses (2) et (3) ni la conclusion, pour se ramener au 1er cas. Il suffit donc de montrer que dans le 1er cas on a $H'' \cong H$.

2). Pour tout $A \subset X$, $|A| = h-1$, il existe $B \subset Y$, $|B| = h''-1$, tel que

$$(4) \qquad E_i \supset A \Longleftrightarrow G_i \supset B \qquad (i \in M)$$

Il suffit de montrer cela lorsque $|A \cap A_o| = h-2$. Soit $x \notin A_o \cup A$ et posons $E_i = A_o \cup A$, $E_j = A_o + x$, $E_k = A+x$. Comme $E_k \not\supset A_o$ on a $G_k \not\supset B_o$ et d'après le lemme $G_k \subset G_i \cup G_j$; posons $B = G_i \cap G_j$. Soit $E_\ell \supset A$ ($\ell \neq i,j,k$), d'après le lemme encore on a $G_\ell \subset G_i \cup G_k$ ou $G_\ell \supset B$. Mais si $G_\ell \subset G_i \cup G_k$ on a $|G_\ell \cap G_j| = h''-1$ qui contredit $|E_\ell \cap E_j| < h-1$. Donc $G_\ell \supset B$. Si $G_\ell \supset B$ on montre de même avec E_i, E_j, E_k à la place de G_i, G_j, G_k que $E_\ell \supset A$.

3). Supposons indexé l'ensemble des $(h-1)$-parties de X :

$$\mathcal{P}_{h-1}(X) = (A_p / p \in N) = L$$

et posons $L' = (B_p / p \in N)$ où B_p est associé à A_p suivant (4).

Alors $L \cong K_n^{h-1}$, L' est $(h''-1)$-uniforme et vérifie :

(2') $\bigcap_{p \in N} B_p = \emptyset$, car chaque B_p est inclus dans un G_i et $\bigcap_{i \in M} G_i = \emptyset$

(3') $|B_p \Delta B_q| = |A_p \Delta A_q|$ $(p,q \in N)$. En effet cela est immédiat à vérifier lorsque $|A_p \Delta A_q| = 2$; soit ensuite une chaîne d'arêtes A_r allant de A_p à A_q telle que $|A_r \Delta A_{r+1}| = 2$; si cette chaîne est prise minimale sa longueur est $\ell = \frac{1}{2} |A_p \Delta A_q|$. On a alors $|B_r \Delta B_{r+1}| = 2$ dans L' et l'existence de cette chaîne de B_r entre B_p et B_q nécessite que $|B_p - B_q| = \frac{1}{2} | B_p \Delta B_q | \leq \ell$ d'où $|B_p \Delta B_q| \leq | A_p \Delta A_q|$. Pour montrer l'inégalité inverse soient x, $x' \notin A_p \cup A_q$, $x \neq x'$, et posons $E_i = A_p + x$, $E_j = A_q + x'$. On a

$$\frac{1}{2} |B_p \Delta B_q| = |B_p - B_q| \geq |G_i - G_j| - 1 = \frac{1}{2} |G_i \Delta G_j| - 1$$

$$= \frac{1}{2} |E_i \Delta E_j| - 1 = \frac{1}{2} |A_p \Delta A_q|$$

3). Par hypothèse de récurrence appliquée à L et L', on a :

- soit $\overline{L} \cong L'$, et alors si φ est la bijection de l'ensemble des sommets de $\overline{L}$, c'est à dire X, sur celui de L' :

$$G_i = \bigcup_{p/A_p \subset E_i} B_p = \bigcup_{p/A_p \subset E_i} \varphi(\overline{A}_p) = \varphi\left(\bigcup_{p/A_p \subset E_i} \overline{A}_p\right) = \varphi\left(\bigcap_{p/A_p \subset E_i} A_p\right) = \varphi(X)$$

ce qui implique que H' n'a qu'une seule arête et donc aussi H, mais cela contredit $n \geqslant 2h$.

- soit $L \cong L'$, et alors

$$G_i = \bigcup_{p/A_p \subset E_i} B_p = \bigcup_{p/A_p \subset E_i} \varphi(A_p) = \varphi\left(\bigcup_{p/A_p \subset E_i} A_p\right) = \varphi(E_i)$$

d'où (par φ) $H'' \cong H$, C.Q.F.D.

REFERENCES

1. C. Berge, Une condition pour qu'un hypergraphe soit fortement isomorphe à un hypergraphe complet ou multiparti, C.R. Acad. Sc. Paris 274 (1972), 1783-1786.

2. W.H. Cunningham, On Theorems of Berge and Fournier, this volume, p.69.

3. J.C. Fournier, Sur les isomorphismes d'hypergraphes, C.R. Acad. Sc. Paris, 274 (1972), 1612-1614.

ON A PROPERTY OF HYPERGRAPHS WITH NO CYCLES OF LENGTH GREATER THAN TWO

P. Hansen, University of Lille
M. Las Vergnas, C.N.R.S.

The result of this note extends a theorem of Lovász [3] on hypergraphs without cycles of length ≥ 3, and was found independently by the two authors (cf. [2], chap. II, section 2, Corollary 2 of Proposition 2). The proof given here is new, and shorter than [3]. The definitions and notations are given in [1].

Lemma. *Every hypergraph without cycles of length greater than 2 has a vertex that belongs to only one edge, or else there exist two edges E_i and E_j with $E_i \subset E_j$.*

Let $H = (X, (E_i)_{i \in I})$ be a hypergraph without cycles of length ≥ 3. We shall assume that no edge contains any other, and that every vertex belongs to at least two edges. Let

$$(x_1, E_{i_1}, x_2, E_{i_2}, \ldots, x_p, E_{i_p}, x_{p+1})$$

be a chain in H of maximal length.

We may assume that $x_1 \in E_{i_1} - E_{i_2}$ because, otherwise, x_1 could be replaced with a vertex x such that $x \in E_{i_1} - E_{i_2}$; this vertex x exists, and x is different from x_k for $k=2,3$ (because $x_2, x_3 \in E_{i_2}$) and for $4 \leq k \leq p+1$ (because there exists no cycle of length ≥ 3).

There exists an edge E_i with $i \neq i_1$ and $x_1 \in E_i$. Since $x_1 \notin E_{i_2}$, we have $i \neq i_2$; furthermore, if $i = i_k$, $3 \leq k \leq p$, there would exist a cycle $(x_1, E_{i_1}, x_2, \ldots, x_k, E_{i_k}, x_1)$ of length ≥ 3, which is a contradiction.

Therefore, by the maximality of the chain $(x_1, \ldots, x_{p+1})$, we have

$$E_i \subset \{x_1, x_2, \ldots, x_{p+1}\}$$

Since $i \neq i_1$, we have $E_i - E_{i_1} \neq \emptyset$; let k be the smallest possible index such that $x_k \in E_i - E_{i_1}$.

We have $k \neq 1,2$, because $x_k \notin E_{i_1}$; we have $k < 3$, because otherwise, there would exist a cycle $(x_1, E_{i_1}, x_2, \ldots, x_k, E_i, x_1)$ of length ≥ 3. The required contradiction follows.

Theorem. Let $H = (X, (E_i)_{i \in I})$ be a hypergraph without cycles of length ≥ 3, with p connected components , such that any two edges have at most s vertices in common. Then

$$(1) \qquad \sum_{i \in I} (|E_i|) - s) \leq |X| - ps$$

The theorem being true for $\Sigma|E_i| = 1$, we shall assume that it is true for all hypergraphs H' with $\Sigma|E_i^!| < \Sigma|E_i|$.

By the lemma, only two cases can occur :

Case 1 : There exists a vertex x_1 that belongs to only one edge, say E_1. Since the theorem is true for the subhypergraph H' induced by $X - \{x_1\} = X'$ by the induction hypothesis, we have :

$$\sum_{i \in I'} (|E_i^!| - s) \leq |X'| - p's$$

If $E_1 \neq \{x_1\}$, then $I' = I$, $p' = p$, $|E_1'| = |E_1|-1$, and (1) follows.

If $E_1 = \{x_1\}$, then $I' = I - \{1\}$, $p' = p-1$, and (1) follows.

Case 2 : There exists no vertex that belongs to only one edge, but there exist two edges E_{i_o} and E_{j_o} with $E_{j_o} \subset E_{i_o}$. Since the theorem is true for the partial hypergraph $H' = (E_i / i \in I - \{j_o\})$, we have

$$\sum_{i \in I-j_o} (|E_i| -s) \leq |X'| - p's$$

Clearly, $X'=X$, and $p'=p$. Furthermore,

$$|E_{j_o}| -s = |E_{i_o} \cap E_{j_o}| - s \leq 0$$

Hence, (1) follows.

Q.E.D.

REFERENCES

1. C. Berge, Graphes et Hypergraphes, Dunod, Paris 1970

2. M. Las Vergnas, Problèmes de couplages et problèmes hamiltoniens en Théorie des Graphes, Thèse, Paris 1972.

3. L. Lovász, Graphs and set-systems, in Beiträge zur Graphentheorie, éd. H. Sachs, H.S. Voss & H. Walther, Teubner, 1968, pp. 99-106.

SUR LES HYPERGRAPHES BICHROMATIQUES

M. Las Vergnas, C.N.R.S.

1. Introduction : rappel de quelques résultats concernant les propriétés de coloration des hypergraphes finis.

Un hypergraphe $H = (X, \mathcal{E})$, $\mathcal{E} = (E_i)_{i \in I}$, est dit <u>équilibré</u> si dans tout cycle $(x_1, E_{i_1}, x_2, \ldots, x_\ell, E_{i_\ell}, x_1)$ de H de longueur ℓ impaire il existe un sommet x_i qui appartient à au moins 3 arêtes du cycle (C. Berge [1], [2]). Les hypergraphes équilibrés généralisent en plusieurs sens les graphes bipartis : en particulier C. Berge a montré qu'un hypergraphe équilibré est bichromatique [2] (le <u>nombre chromatique</u> $\chi(H)$ d'un hypergraphe $H = (X, \mathcal{E})$ est le plus petit entier k tel qu'il existe une partition $X = S_1 + S_2 + \ldots + S_k$ de X avec la propriété qu'aucun des S_i ne contient une arête de cardinal ≥ 2 de H ; H est <u>bichromatique</u> si $\chi(H) \leq 2$).

L. Lovász dans [8] introduit la notion d'hypergraphe <u>normal</u> : un hypergraphe H est normal si $\nu(H') = \tau(H')$ pour tout hypergraphe partiel H' de H (soit H un hypergraphe : $\nu(H)$ est le cardinal maximal d'un couplage de H i.e. d'un ensemble d'arêtes 2 à 2 disjointes, $\tau(H)$ est la cardinal minimal d'un transversal de H i.e. d'un ensemble de sommets rencontrant toutes les arêtes de H ; on a clairement $\nu(H) \leq \tau(H)$).

Théorème (C. Berge & M. Las Vergnas [3]) : un hypergraphe est équilibré si et seulement si $\nu(H') = \tau(H')$ pour tout sous-hypergraphe partiel H' de H. Un hypergraphe équilibré est donc normal. L. Lovász dans une première version non publiée de [8] posait la question : un hypergraphe normal est-il nécessairement bichromatique ?

La réponse (affirmative) fut obtenue indépendamment par J.-C. Fournier et l'auteur sous la même forme suivante : disons qu'un hypergraphe H est <u>pseudo-équilibré</u> si dans tout cycle de H de longueur impaire il existe 3 arêtes qui ont une intersection non vide (ou encore, de façon équivalente si H ne contient pas de cycle de longueur impaire de degré maximal 2), un hypergraphe normal est pseudo-équilibré car un cycle H' de longueur impaire 2p+1 s'il est de degré maximal 2 est tel que $\nu(H') \leq p < \tau(H')$ (notons que par contre un hypergraphe pseudo-équilibré n'est pas nécessairement normal ; exemple l'hypergraphe sur l'ensemble 1234567 ayant pour arêtes 123,234,345,456,567,167,127 pseudo-équilibré mais non normal),

Théorème A (J.-C. Fournier & M. Las Vergnas [5]) : un hypergraphe pseudo-équilibré est bichromatique.

2. Nous considérerons désormais des hypergraphes quelconques (finis ou infinis) ; les définitions des hypergraphes bichromatiques et pseudo-équilibrés données au §1 s'étendent sans changement. Sans perdre en généralité on peut dans ce qui suit se restreindre à ne considérer que des hypergraphes sans arêtes multiples. Nous conviendrons dès lors pour simplifier les notations que dans l'écriture $H = (X, \mathcal{E})$ $\mathcal{E}$, l'ensemble des arêtes de H , est un sous-ensemble de l'ensemble des parties de X . Pour $E \in \mathcal{E}$ H-E désignera l'hypergraphe partiel de H ayant $\mathcal{E} - \{E\}$ pour ensemble d'arêtes.

Nous dirons qu'un hypergraphe $H = (X, \mathcal{E})$ est une <u>obstruction au bichromatisme</u> si 1°) $\chi(H) = 3$ et 2°) $\chi(H-E) = 2$ pour tout $E \in \mathcal{E}$. Clairement un hypergraphe fini est bichromatique si et seulement si aucun de ses hypergraphes partiels n'est une obstruction au bichromatisme. Le théorème A est équivalent à ce que dans toute obstruction au bichromatisme finie il existe un cycle de longueur impaire de degré maximal 2 .

Lemme [5] : tout cycle de longueur impaire de degré maximal 2 contient (i.e. a un hypergraphe partiel qui est) un cycle de longueur impaire de degré maximal 2 tel que 2 arêtes quelconques de ce cycle non-consécutives aient une intersection vide (n.b. pour un cycle de longueur ≥ 4 cette dernière propriété entraîne que le degré maximal est 2). En conséquence le théorème A est encore équivalent à ce que dans toute obstruction au bichromatisme finie il existe un cycle de longueur impaire de degré maximal 2 tel que 2 arêtes non-consécutives aient une intersection vide.

Proposition 1.

Soit $H = (X, \mathcal{E})$ un hypergraphe et $E_o \in \mathcal{E}$. Supposons que $\chi(H-E_o) = 2$: alors s'il existe $x_o \in E_o$ avec la propriété que H ne contienne aucun cycle de longueur impaire de degré maximal 2 tel que 2 arêtes non-consécutives aient une intersection vide, de la forme $(x_1, E_1 = E_o, x_2 = x_o, \ldots, x_\ell, E_\ell, x_1)$ avec $E_1 \cap E_2 = \{x_o\}$, on a $\chi(H) = 2$.

Démonstration.

Considérons une bicoloration $X = S+T$ de $H-E_o$, les notations étant supposées telles que $E_o \subset S$ (si E_o rencontre S et T il n'y a rien à démontrer). Posons $\mathcal{E}_o = \{E_o\}$, $\mathcal{F}_o = \emptyset$, $A_o = \{x_o\}$, $B_o = \emptyset$. Soit $\mu \geq 1$ un ordinal et supposons que pour tout ordinal $\lambda < \mu$ on ait défini $A_\lambda \subset S$, $B_\lambda \subset T$: nous posons

$$S_\mu = S - \bigcup_{\lambda<\mu} A_\lambda + \bigcup_{\lambda<\mu} B_\lambda \, , \quad T_\mu = T + \bigcup_{\lambda<\mu} A_\lambda - \bigcup_{\lambda<\mu} B_\lambda \, ,$$

$$\mathcal{E}_\mu = \{E \in \mathcal{E} \text{ t.q. } E \subset S_\mu\} \, , \quad \mathcal{F}_\mu = \{E \in \mathcal{E} \text{ t.q. } E \subset T_\mu\},$$

$$A_\mu = \bigcup_{E \in \mathcal{E}_\mu} (E \cap S) \, , \quad B_\mu = \bigcup_{E \in \mathcal{F}_\mu} (E \cap T) \, .$$

Je dis que pour $\mu \geq 1$ $E_o \notin \mathcal{E}_\mu + \mathcal{F}_\mu$ et que les A_μ, B_μ, $\mu = 0,1,\ldots$ sont 2 à 2 disjoints. La proposition en résulte. En effet par suite il existe un ordinal μ_o tel que $A_{\mu_o} = B_{\mu_o} = \emptyset$: pour tout $E \in \mathcal{E}_{\mu_o}$ (resp. $\mathcal{F}_{\mu_o}$) on a $E \cap S = \emptyset$ (resp. $E \cap T = \emptyset$) d'où puisque toute arête de H différente de E_o rencontre S et T et que $E_o \notin \mathcal{E}_{\mu_o} + \mathcal{F}_{\mu_o}$ $\mathcal{E}_{\mu_o} + \mathcal{F}_{\mu_o} = \emptyset$ i.e. $X = S_{\mu_o} + T_{\mu_o}$ est une bicoloration de H.

Nous montrons par récurrence sur μ (récurrence transfinie, cf [4]) que pour $\mu \geq 1$ 1°) $E \notin \mathcal{E}_\mu + \mathcal{F}_\mu$, 2°) $A_\mu \cap A_\lambda = B_\mu \cap B_\lambda = \emptyset$ pour $\lambda < \mu$, et 3°) $E_o \cap A_\mu = \emptyset$ (par définition $E_o \cap B_\mu = \emptyset$). Soit donc μ un ordinal >1 , supposons l'hypothèse de récurrence vérifiée pour tout ordinal λ tel que $1 \leq \lambda < \mu$ et vérifions-la pour μ (la vérification pour $\mu = 1$ est immédiate):

E_o est bicoloré par $(S-A_o+B_o, T+A_o-B_o)$, ne rencontrant pas A_λ, B_λ pour $1 \leq \lambda < \mu$ E_o est bicoloré par (S_μ, T_μ) : $E_o \notin \mathcal{E}_\mu + \mathcal{F}_\mu$. Supposons $A_\mu \cap A_\lambda \neq \emptyset$ pour un ordinal λ $0 \leq \lambda < \mu$: il existe $E \in \mathcal{E}_\mu$ tel que $E \cap A_\lambda \neq \emptyset$, alors $E \cap T_\mu \neq \emptyset$ ce qui contredit $E \subset S_\mu$ (de même $B_\mu \cap B_\lambda = \emptyset$ pour $\lambda < \mu$).

Soit $E \in \mathcal{E}_\mu$; construisons une chaîne ayant pour arêtes successives $E_1=E, E_2, \ldots, E_\ell=E_o$ de la façon suivante : on prend $E_1 = E$; $E_1 \subset S_{\lambda_1}$,où $\lambda_1 = \mu$, d'où $E_1 \cap T \subset \sum_{\lambda < \lambda_1} B_\lambda$, $E_1 \cap T \neq \emptyset$ (car $E_o \notin \mathcal{E}_{\lambda_1}$) soit λ_2 le plus petit ordinal λ tel que $E_1 \cap B_\lambda \neq \emptyset$, on prend $E_2 \in \mathcal{F}_{\lambda_2}$ tel que $E_2 \cap E_1 \neq \emptyset$; $E_2 \subset T_{\lambda_2}$ d'où $E_2 \cap S \subset \sum_{\lambda < \lambda_2} A_\lambda$, $E_2 \cap S \neq \emptyset$ soit λ_3 le plus petit ordinal λ tel que $E_2 \cap A_\lambda \neq \emptyset$, on prend $E_3 \in \mathcal{E}_{\lambda_3}$ tel que $E_3 \cap E_2 \neq \emptyset$. Si $\lambda_3 = 0$ $E_3 = E_o$ la construction est terminée. Si $\lambda_3 \geq 1$ on a $E_3 \cap T \subset \sum_{\lambda < \lambda_3} B_\lambda$, $E_3 \neq E_o$ d'où $E_3 \cap T \neq \emptyset$, soit λ_4 le plus petit ordinal λ tel que $E_3 \cap B_\lambda \neq \emptyset$, etc... La suite d'ordinaux $\lambda_1 > \lambda_2 > \lambda_3 > \ldots$ ainsi construite étant strictement décroissante est finie : elle se termine nécessairement pour un entier ℓ par $\lambda_\ell = 0$ et $E_\ell = E_o$. E_i appartenant à $\mathcal{E}_{\lambda_i}$ si i est impair, à $\mathcal{F}_{\lambda_i}$ si i est pair ℓ est impair. $E_i \subset A_{\lambda_i} + \sum_{\lambda_{i+1} \leq \lambda < \lambda_i} B_\lambda$ pour i impair $(=1,3,\ldots,\ell-2)$, $E_i \subset B_{\lambda_i} + \sum_{\lambda_{i+1} \leq \lambda < \lambda_i} A_\lambda$ pour i pair $(=2,\ldots,\ell-1)$: 2 arêtes non consécutives dans la suite $E_1, E_2, \ldots, E_\ell$ ont une intersection vide sauf peut-être $E_1 = E$ et $E_\ell = E_o$. Cependant si l'on avait $E \cap E_o \neq \emptyset$ (alors $E \cap E_o \subset E_o - x_o$ car $E \cap A_o = \emptyset$) les arêtes $E_o, E_1, E_2, \ldots, E_\ell = E_o$ dans cet ordre constitueraient un cycle contredisant l'hypothèse de la proposition. Ainsi $E \in \mathcal{E}_\mu$ entraîne $E \cap E_o = \emptyset$ d'où $A_\mu \cap E_o = \emptyset$.

<u>Remarque</u> : $A_n = \emptyset$ pour n entier impair (resp. $B_n = \emptyset$ pour n entier

pair) : dans le cas fini la démonstration est celle du théorème A donnée dans [7] chap. II §3 ; par contre pour $\lambda \geqslant \omega$ (ordinal de l'ensemble des entiers) A_λ et B_λ peuvent être simultanément non vides. On peut montrer en outre sans difficulté que pour $E \in \mathcal{E}_\mu$ on a $E \cap B_{\mu'} \neq \emptyset$ si $\mu = \mu'+1$ (resp. pour tout $\lambda < \mu$ il existe μ' $\lambda < \mu' < \mu$ tel que $E \cap B_{\mu'} \neq \emptyset$ si μ est sans prédecesseur) - on sait dans la démonstration que $\emptyset \neq E \cap T \subset \sum_{\lambda<\mu} B_\lambda$ -, propriété analogue pour $E \in \mathcal{F}_\mu$: par suite les $\mathcal{E}_\mu$, $\mathcal{F}_\mu$ $\mu = 0,1,\ldots$ sont 2 à 2 disjoints.

Proposition 2.

Soit $H = (X, \mathcal{E})$ une obstruction au bichromatisme : pour tout $E \in \mathcal{E}$ et $x \in E$ il existe un cycle de longueur impaire et de degré maximal 2 tel que 2 arêtes non consécutives aient une intersection vide, de la forme $(x_1, E_1 = E, x_2 = x, \ldots, x_\ell, E_\ell, x_1)$ avec $E_1 \cap E_2 = \{x\}$.

Conséquence immédiate de la proposition 1 .

Lemme.

Soit H un hypergraphe dont toutes les arêtes sont finies : si tous les hypergraphes partiels de H comportant un nombre fini d'arêtes sont bichromatiques alors H est bichromatique.

Démonstration.

Ce lemme est une application du théorème 1 §7 de [4] sur les limites projectives, "bichromatique" pouvant évidemment être remplacé par "k-chromatique" pour k entier quelconque. Soit $H = (X, \mathcal{E})$: pour $\mathcal{F} \subset \mathcal{E}$ posons $X_{\mathcal{F}} = \bigcup_{E \in \mathcal{F}} E$ et soit $C(\mathcal{F})$ l'ensemble des bicolorations de l'hypergraphe $(X_{\mathcal{F}}, \mathcal{F})$ considérées par exemple comme applications de X dans $\{1,2\}$. Pour $\mathcal{F} \subset \mathcal{F}'$ l'application qui à une bicoloration appartenant à $C(\mathcal{F}')$ fait correspondre sa restriction à $X_{\mathcal{F}}$ est une application de $C(\mathcal{F}')$ dans $C(\mathcal{F})$. On vérifie immédiatement que les $C(\mathcal{F})$, $\mathcal{F}$ fini $\subset \mathcal{E}$, munis de ces applications constituent un système projectif et que

$\lim_{\mathcal{F} \text{ fini} \subset \mathcal{E}} C(\mathcal{F})$ est l'ensemble $C(\mathcal{E})$ des bicolorations de H . Pour $\mathcal{F}$ fini $\subset \mathcal{E}$ $C(\mathcal{F})$ est fini (car $X_{\mathcal{F}}$ est fini) et non vide par hypothèse : par le thèorème de [4] $C(\mathcal{E})$ est non vide.

Proposition 3.

Un hypergraphe pseudo-équilibré ne comportant qu'un nombre fini d'arêtes de cardinal infini est bichromatique.

Application du lemme et de la proposition 1 .

Il existe des hypergraphes bichromatiques qui ne sont pas pseudo-équilibrés (exemples : les hypergraphes sur l'ensemble 1234 H_1 ayant pour arêtes 12, 13, 234 , H_2 ayant pour arêtes 12, 134, 234). Soit $H = (X, \mathcal{E})$ un hypergraphe : considérons la relation R sur X définie par xRy si et seulement si $x \in E$ est équivalent à $y \in E$ pour tout $E \in \mathcal{E}$. R est une relation d'équivalence. Soit $\hat{X}$ un ensemble de représentants des classes d'équivalence dont les sommets appartiennent à au moins 2 arêtes de H . Nous noterons $\hat{H}$ le sous-hypergraphe de H engendré par $\hat{X}$ (les hypergraphes $\hat{H}$ correspondant à différents $\hat{X}$ sont isomorphes). L'opération $\hat{}$ permet d'énoncer une réciproque à la proposition 3 analogue à celle donnée par C. Berge dans le cas des hypergraphes équilibrés (un hypergraphe est équilibré si et seulement si tous ses sous-hypergraphes partiels sont bichromatiques [2]):

Proposition 4.

Soit H un hypergraphe ne comportant qu'un nombre fini d'arêtes de cardinal infini : H est pseudo-équilibré si et seulement si $\hat{H'}$ est bichromatique pour tout hypergraphe partiel H' de H .

La démonstration à partir de la proposition 3 et du lemme de [5] (cf §1) consiste dans les 2 remarques suivantes :

1) soit H un hypergraphe pseudo-équilibré, alors $\hat{H}$ est pseudo-équilibré,

2) si H' est un cycle de longueur impaire et de degré maximal 2 tel que 2 arêtes non-consécutives aient une intersection vide alors $\hat{H'}$ est un

cycle de longueur impaire uniforme de rang 2 : $\chi(H') = 3$.

Notons que si pour un hypergraphe H $\hat{H}$ est bichromatique H est lui-même bichromatique (H_1 et H_2 étant évidemment des contre-exemples à la réciproque): la proposition 4 contient la proposition 3 .

3. Dans toute obstruction au bichromatisme il existe un cycle de longueur impaire de degré maximal 2 tel que 2 arêtes quelconques non-consécutives aient une intersection vide. Ce résultat est contenu dans la conjecture suivante proposée par F. Sterboul pour un hypergraphe fini :

Conjecture (F. Sterboul).

Dans toute obstruction au bichromatisme il existe un cycle de longueur impaire tel que

(i) l'intersection de 2 arêtes quelconques non-consécutives soit vide

et (ii) l'intersection de 2 arêtes consécutives quelconques est de cardinal 1 sauf peut-être pour un couple d'arêtes consécutives.

(n.b. les propriétés (i) et (ii) entrainent que le degré maximal du cycle est 2)

La restriction "sauf peut-être pour un couple " ne peut être levée (contre-exemple : l'hypergraphe ayant pour arêtes les parties à n+1 éléments d'un ensemble de cardinal 2n+1 , $n \geqslant 2$, obstruction au bichromatisme, ne contient pas de cycle de longueur impaire de degré maximal 2 tel que l'intersection de 2 arêtes consécutives quelconques soit de cardinal 1). D'autre part considérons l'hypergraphe H sur l'ensemble 123456789 ayant pour arêtes 12,235,34,45,356,679,78,89,179 : on vérifie que H est une obstruction au bichromatisme et qu'il n'existe aucun cycle avec les propriétés (i) et (ii) de la conjecture contenant l'arête 12 , la conjecture ne peut pas être renforcée de façon à contenir la proposition 2 .

Soit H une obstruction au bichroma tisme : pour toute arête E de H et tout $x \in E$ il existe une arête E' de H telle que $E \cap E' = \{x\}$ (soit X = S+T une bicoloration de H-E , les notations étant supposées telles que $E \subset S$; il existe une arête E' contenue dans T+x sinon (S-x,T+x) serait une bicoloration de H : $\emptyset \neq E' \cap S = \{x\} = E \cap E'$). A partir de cette remarque il est aisé de construire lorsque H est fini un cycle de H avec les propriétés (i) et (ii) de la conjecture : soient E_1 une arête de H , $x_2 \in E_1$, E_2 tel que $E_2 \cap E_1 = \{x_2\}$, $x_3 \in E_2 - E_1$ (les arêtes d'une obstruction au bichromatisme sont 2 à 2 incomparables pour l'inclusion), il existe une arête E_3 telle que $E_3 \cap E_2 = \{x_3\}$, si $E_3 \cap E_1 \neq \emptyset$ soit $x_1 \in E_3 - E_1$ $(x_1, E_1, x_2, E_2, x_3, E_3, x_1)$ est le cycle cherché, sinon soit $x_4 \in E_3 - E_2$, etc.... Soit ℓ $(\geqslant 3)$ le plus petit indice tel que E_ℓ ait une intersection non vide avec l'une des arêtes $E_1, E_2, \ldots, E_{\ell-2}$ et soient k le plus grand indice $1 \leqslant k < \ell-2$ tel que $E_k \cap E_\ell \neq \emptyset$ et $x \in E_k \cap E_\ell$: le cycle $(x, E_k, x_{k+1}, \ldots, x_\ell, E_\ell, x)$ a les propriétés (i) et (ii) de la conjecture. Le problème est de montrer qu'il existe un tel cycle de longueur impaire.

<u>Note</u> : J.-C. Fournier et l'auteur ont démontré la version affaiblie suivante de la conjecture, intermédiaire entre celle-ci et le théorème A [6] :

dans toute obstruction au bichromatisme il existe un (et en fait 2 si l'obstruction est différente d'un cycle de longueur impaire uniforme de rang 2) cycle de longueur impaire de degré maximal 2 tel que sauf peut-être pour un couple, 2 arêtes consécutives aient une intersection de cardinal 1 .

Ce résultat entraîne la conjecture dans le cas des hypergraphes uniformes de rang $\leqslant 3$.

REFERENCES

1. C. BERGE , <u>Graphes et Hypergraphes</u>, Dunod, Paris 1970
2. C. BERGE , Sur certains hypergraphes généralisant les graphes bipartites, pp 119-133 in <u>Combinatorial Theory & its Applications</u>, Proc. Colloq. Balatonfüred, Amsterdam 1970
3. C. BERGE & M. LAS VERGNAS , Sur un théorème de type König pour hypergraphes, <u>Ann. New-York Acad. Sci.</u>, <u>175</u> (1970), 19-31
4. N. BOURBAKI , <u>Théorie des Ensembles</u>, chap. III , 2ème éd., Hermann, Paris 1963

5. J.-C. FOURNIER & M. LAS VERGNAS , Une classe d'hypergraphes bichromatiques, Discrete Math., 2 (1972), 407-410

6. J.-C. FOURNIER & M. LAS VERGNAS , Une classe d'hypergraphes bichromatiques (II), à paraître à Discrete Math.

7. M. LAS VERGNAS , Problèmes de couplages et problèmes hamiltoniens en Théorie des Graphes, Thèse, Paris 1972 (non publié)

8. L. LOVASZ , Normal hypergraphs and the Perfect Graph Conjecture, Discrete Math., 2 (1972),253-268

MINIMAX THEOREMS FOR HYPERGRAPHS

L. Lovász, Vanderbilt University

I. Introduction. Obviously, the chromatic number $\chi(G)$ of a graph G is at least as large as the maximum number $\omega(G)$ of vertices in a clique of G. Generally, $\chi(G)$ is much larger than $\omega(G)$; there are graphs without triangles with arbitrarily high chromatic number. However, there are very interesting classes of graphs for which $\chi(G) = \omega(G)$; like bipartite graphs (trivially), complements of bipartite graphs (by a theorem of König), line-graphs of bipartite graphs (by another theorem of König), complements of line graphs of bipartite graphs (that is equivalent to the Frobenius-König-Hall theorem), interval graphs (by a theorem of Gallai) transitively orientable graphs (by a theorem of Erdös-Szekeres) complements of transitively orientable graphs (by a theorem of Dilworth) etc. All these examples have the property that all of their induced subgraphs are graphs of similar type, thus satisfying $\chi(G) = \omega(G)$. This suggests the following definition, given by Berge:

A graph G is perfect if $\chi(G') = \omega(G')$ for every induced subgraph G' of G.

On the basis of the examples listed above, Berge [1] formulated the following conjecture:

(A) The complement of a perfect graph is perfect.

An odd circuit without diagonals is not perfect; neither is its complement. Therefore, a perfect graph cannot contain an odd circuit without diagonals nor the complement of it. The second conjecture of Berge, still unsettled, states that this property characterizes perfect graphs:

(B) A graph is perfect iff neither itself nor its complement contain odd circuits without chords.[1]

Berge [2] was the first to apply hypergraphs in examination of perfect graphs. Fulkerson [7] found a different (but related) method to prove a theorem slightly weaker than (A); in fact, to deduce (A) from his result only Lemma 1 is needed. In [10], conjecture (A) was proved. Using this result, in [11] it was shown that

(AB) A graph G is perfect iff $\omega(G') \cdot \omega(\overline{G}') \geq n(G')$ holds for every induced subgraph G' of G.

(Here $n(G')$ denotes the number of vertices of G', $\overline{G}'$ is its complement). (AB) expresses that this theorem is stronger than (A) but weaker than (B).

Although not everywhere formulated this way, the proofs of the two latter results use ideas from hypergraph theory and their most natural form is "hypergraphic"; this form leads to some generalizations discussed in this paper.

Given a hypergraph H, we have the evident inequality

$$\nu(H) \leq \tau(H)$$

between the maximum number $\nu(H)$ of disjoint edges of H and the minimum number $\tau(H)$ of points representing all edges. For graphs, the equality holds if H is bipartite and this can be converted in the following sense; if $\nu(G') = \tau(G')$ for each subgraph (partial graph) G' of G, then G has no odd circuits and thus it is bipartite. This raises the problem to characterise a reasonable class of hypergraphs with $\nu(H) = \tau(H)$. Berge [2] and Berge - Las Vergnas [4] gave many interesting properties and manysided characterizations of those hypergraphs--called balanced--which have $\nu(H') = \tau(H')$ for every partial subhypergraph H'. In the

---

[1] See T. Gallai [8], Sachs [12].

present paper we are going to survey those normal hypergraphs H which satisfy $\nu(H') = \tau(H')$ for every partial hypergraph H'. This class is wider than the class of all balanced hypergraphs, as shown by simple examples. It is not so well-characterized, but seems to be very important because of the following fact: there is a simple correspondence between normal hypergraphs and perfect graphs. We shall give several characterizing properties of normal hypergraphs.

A further weakening of this notion is the class of seminormal hypergraphs, characterized by the property that $\nu(H') = \tau(H')$ holds for every section hypergraph H' of them. An interesting example of such hypergraphs was given by J. Edmonds [5]: consider the edges of a directed graph G as points of H, select a "root" $x_0 \in V(G)$ and consider the branchings[2] rooted in x_0 as edges of H. The fact that these hypergraphs are seminormal, is equivalent to his theorem that the maximum number of edge-disjoint branchings rooted in x_0 equals the minimum cardinality of a cut[3]. Another example arises from Menger's theorem: Consider two non-adjacent vertices of a (non-directed) graph G, let $V(H) = V(G) - \{a, b\}$ and consider as edges of H the sets of vertices of paths connecting a to b. We shall give a characterization of semi-normal hypergraphs which, however, cannot be considered as a final one, since it is not evident at all that it contains the mentioned examples. We hope there is a more useful characterization implying Edmonds' and Menger's theorem.

Many considerations of the present paper are related to Fulkerson's results in [7], but discussed from a different point of view.

II. Definitions. A hypergraph H is a non-empty finite collection of non-empty finite sets called edges; we allow multiple edges, i.e., edges which are equal as sets. The elements of the edges are the vertices; their set is V(H). A

---

[2] A branching is a spanning tree with all the edges oriented away from the root.

[3] A cut is the set of edges connecting a set $S \subseteq V(G)$, $x_0 \in S$, to $V(G) - S$.

partial hypergraph is a subfamily of the edges. The subhypergraph generated by $S \subseteq V(H)$ is the hypergraph $H_S = \{S \cap E;\ E \in H,\ E \cap S \neq \emptyset\}$. The section hypergraph induced by a set $S \subseteq V(G)$ is the hypergraph $H \times S = \{E \in H;\ E \subseteq S\}$.

Moreover, we use the following notations:

$n(H)$ = number of vertices in H;

$m(H)$ = number of edges in H;

$\tau(H)$ = minimum number of points representing all edges;

$\nu(H)$ = maximum number of pairwise disjoint edges;

$\delta(H)$ = maximum degree = maximum number of edges with non-empty intersection;

$q(H)$ = chromatic index = minimum number of colors to color the edges in such a way that the edges of the same color are pairwise disjoint.

A hypergraph H is normal if $\tau(H') = \nu(H')$ for every partial hypergraph H' of H; H is q-normal if $q(H') = \delta(H')$ for every partial hypergraph $H' \subseteq H$. However, this notion is only temporary because of the following result:

Theorem 1. A hypergraph is normal if it is q-normal.

The proof of this theorem will be given later. Let us show the connection between normal hypergraphs and perfect graphs. We have to introduce a few notions. Let $L(H)$ be the line-graph of the hypergraph of H, i.e., the graph whose vertices are the edges of H and two of them are joined iff they meet. Let G be a graph; then we define $C(G)$ as the hypergraph whose vertices are the cliques in G, and for any point x of G, there is an edge E_x of $C(G)$ containing all points of $C(G)$, i.e., all cliques of G which cover x. H is said to have the Helly property if any pairwise intersecting family of its edges have a common point. Obviously, both normal and q-normal hypergraphs have the Helly-property. It is also easily seen that $C(G)$ has the Helly-property. It requires repeated application of the definitions only to check

Proposition 1. A graph G is perfect iff $C(G)$ is q-normal; the complement $\overline{G}$ is perfect iff $C(G)$ is normal. A hypergraph H is q-normal iff H has the

Helly property and $L(H)$ is perfect; H is normal iff it has the Helly property and $\overline{L(H)}$ is perfect.

Now Theorem 1 implies (A): for if G is perfect, $C(G)$ is q-normal; hence $C(G)$ is normal and thus $\overline{G} \cong L(C(G))$ is perfect.

The following theorem gives a characterization of normal hypergraphs:

Theorem 2. A hypergraph H is normal iff (1) $\nu(H') \cdot \delta(H') \geq |H'|$ holds for every partial hypergraph H' of H.

It is easily seen that (1) implies the Helly-property. This theorem easily implies Theorem 1. However, Theorem 1 will be used in its proof. Theorem 2 was formulated and proved in terms of graphs in [11]. The proof given here is slightly different; it will follow from a sequence of related results.

II. S - Coverings. An s-covering is a function $p: V(H) \to \{0,1,..s\}$ such that

$$\sum_{i \in E} p(i) \geq s \tag{1}$$

for every edge $E \in H$. The s-covering number is $\tau_s(H) = \min_p \sum_{i=1}^{n} p(i)$ taken for every s-covering p. For $s = 1$, we may consider values 0, 1 only; therefore, $\tau_1(H) = \tau(H)$. Moreover, we have the following inequalities:

$$\tau_s(H) \leq s \cdot \tau_1(H) \tag{2}$$

$$\tau_{s+r}(H) \leq \tau_s(H) + \tau_r(H) \tag{3}$$

$$\tau_s(H) \geq s \cdot \nu(H) \tag{4}$$

This immediately implies:

Proposition 2. $\nu(H) \leq \frac{\tau_s(H)}{s} \leq \tau(H)$; thus if $\nu(H) = \tau(H)$ then $\frac{\tau_s(H)}{s}$ is constant.

Moreover, this implies, by a theorem of Fekete, the following:

Proposition 3. $\tau^*(H) = \lim_{s\to\infty} \frac{\tau_s(H)}{s}$ exists, and is $\leqq \frac{\tau(s)}{s}$ for each s.

To get more information on this $\tau^*(H)$, consider the following linear progamming exercise:

$$(5)\qquad \begin{array}{l} \sum_{i\in E} x_i \geqq 1 \quad (E\in H) \\ x_i \geqq 0 \qquad\qquad (i = 1,2,\dots,n) \\ \hline \text{minimize } \sum_{i=1}^{n} x_i \end{array}$$

Then the minimum is $\tau^*(H)$. In fact, (5) has a rational extremum; let $\pi(i)$ be this value of x_i. Let q be the least common denominator of the rationals $\pi(i)$, and let $s = q$. Then

$$p(i) = s\cdot\pi(i)$$

is an s-covering and conversely, if $p(i)$ is an s-covering then $x_i = \frac{p(i)}{s}$ satisfies the inequalities of (5). Hence, the optimum is $\frac{\tau_s(H)}{s}$ for these values of s. This proves the assertion.

Analogously to s-coverings, we define an s-matching as a function $f\colon H \to \{0,1,\dots s\}$ such that

$$\sum_{E\ni x} f(E) \leq s$$

Consider the number

$$\nu_s(H) = \max_f \sum_E f(E).$$

Then we have

$$\nu_s(H) \geqq s\cdot\nu(H),$$

$$\nu_{s+r}(H) \geqq \nu_s(H) + \nu_r(H), \quad \nu_s(H) \leqq s\cdot\tau(H);$$

The duality theorem of linear programming gives

Proposition 4. $\lim \frac{\nu_s(H)}{s} = \sup \frac{\nu_s(H)}{s} = \tau^*(H)$, and if $\nu(H) = \tau(H)$ then $\nu_s(H) = s\ \tau^*(H)$ for any s.

We would like to find converses to Propositions 3 and 4. For graphs, we have the following very simple statement:

Theorem 3. If G is a graph and $\tau_s(G) = s\ \tau(G)$ for some $s > 1$, then $\nu(G) = \tau(G)$.

Proof. By (2) and (3), it follows that $\tau_2(T) = 2\ \tau(T)$. Consider a set T representing every edge, $|T| = \tau(G)$. We claim there is a matching which associates a point of $V(G) - T$ with every point of T. By the König-Hall theorem it is enough to prove that the set A of points of $V(G) - T$, connected to points of a given $B \subseteq T$, has cardinality $|A| \geq |B|$. Assume this is not the case. Then set

$$p(x) = \begin{cases} 2, & \text{if } x \in T - B \\ 1, & \text{if } x \in A \cup B \\ 0 & \text{otherwise.} \end{cases}$$

Then $p(x)$ is a 2-cover and

$$\sum_x p(x) = 2|T-B| + |A \cup B| = 2|T| + |A| - |B| < 2|T|,$$

a contradiction.

Q.E.D.

For hypergraphs, there is an example which disproves any kind of simple converse to Proposition 2: let $x_1,\dots,x_6$ be the points of H and $\{x_1,x_2,x_3\}$, $\{x_1,x_4,x_5\}$, $\{x_2,x_5,x_6\}$, $\{x_3,x_4,x_6\}$ be the edges. Then, simply summarizing the inequalities in (1), we get

$$2\Sigma p(x_i) \geqq 4s,$$

i.e., $\tau_s(H) \geqq 2s$ for any s. Since $\tau(H) = 2$, we have $\frac{\tau_s(H)}{s} = 2$ for any s. On the other hand, $\nu(H) = 1$.

However, if we consider all partial hypergraphs of H, we give two characterizations of normal hypergraphs:

Theorem 4. A hypergraph H is normal iff $\tau_2(H') = 2\tau(H')$ holds for each partial hypergraph H' of H.

Theorem 5. A hypergraph H is normal iff $\tau^*(H')$ is an integer for each partial hypergraph H' of H.

For ν, we have a statement of a somewhat stronger type:

Theorem 6. A hypergraph is seminormal iff $\nu^*(H') = \nu(H')$ for every section hypergraph of it.

Before proving these theorems, we introduce the following notions: A hypergraph H is τ-critical,[4] if $\tau(H') < \tau(H)$ for any proper partial hypergraph H' of H. H is ν-critical, if $\nu(H-H_x) = \nu(H)$ for any x (the word critical is used for graphs of this property by Gallai [9]; it refers to the fact that for such connected graphs, $H - H_x$ has a 1-factor for any x). H is hypercritical if every proper partial hypergraph of it is normal, but H itself is not. A τ-critical hypergraph is called trivial if it consists of independent edges.

From these definitions, we easily deduce the following properties:

(i) H is normal if and only if it contains no non-trivial τ-critical partial hypergraph.

(ii) H is normal iff it contains no ν-critical partial hypergraphs.

(iii) H is normal iff it contains no hypercritical partial hypergraphs,

(iv) H is seminormal iff it contains no ν-critical induced partial hypergraphs.

---

[4] For some other properties of τ-critical hypergraphs, see [3], Chap. 18.

(v) a hypercritical hypergraph is τ-critical and ν-critical.

(vi) a hypergraph is hypercritical if and only if it is a nontrivial τ-critical hypergraph and contains no other non-trivial τ-critical hypergraph.

Remark: The converse of (v) is not true. The hypergraph consisting of all triples of neighboring points of a cycle of length 10 is τ-critical and ν-critical, but not hypercritical.

Theorem 4 follows from (i) and the following

Theorem 7. Let H be a τ-critical hypergraph and $s \leq \delta(H)$. Then

$$\tau_s(H) \leq s \cdot \tau(H) - s + 1$$

Proof: Let $E_1,\dots,E_s$ be edges with a common point x^0 (by $s \leq \delta(H)$ such edges exist). Let $p_i(x)$ be a minimum 1-covering of $H - \{E_i\}$; obviously, $p_i(x_0)=0$. Put

$$p(x) = \begin{cases} \sum_{i=1}^{s} p_i(x) & \text{if } x \neq x_0 \\ 1 & \text{if } x = x_0 \end{cases}$$

Then $p(x)$ is an s-covering. For if $E \neq E_i$, then

$$\sum_{x \in E} p(x) \geq \sum_{i=1}^{s} \sum_{x \in E} p_i(x) \geq s,$$

and for $E = E_1$ (say) we have

$$\sum_{x \in E_1} p(x) \geq 1 + \sum_{i=2}^{s} \sum_{x \in E} p_i(x) \geq 1 + (s-1) = s.$$

On the other hand, we have

$$\sum_x p(x) = 1 + \sum_{i=1}^{s} \sum_x p_i(x) = 1 + s(\tau(H) - 1) = s\tau(H) - s + 1,$$

which proves the theorem.

By help of (ii), we can prove Theorem 6 in the following stronger form:

<u>Theorem 8</u>: If H is ν-critical and $s = |E_0|$ for some $E_0 \in H$ then

$$\nu_s(H) \geqq s \quad \nu(H) + 1$$

<u>Proof</u>: Consider a maximal 1-matching $f_x(E)$ of $H - H_x$ for every $x \in E_0$, and put

$$f(E) = \begin{cases} \sum_{x \in E_0} f_x(E) & \text{if } E \neq E_0 \\ 1 & \text{if } E = E_0 . \end{cases}$$

Then $f(E)$ is an s-matching since

$$\sum_{\substack{E \\ y \in E}} f(E) = \sum_{x \in E_0} \sum_{\substack{E \\ y \in E}} f_x(E) \leqq |E_0| = s$$

if $y \notin E_0$ and

$$\sum_{\substack{E \\ y \in E}} f(E) = 1 + \sum_{\substack{x \in E_0 \\ x \neq y}} \sum_{\substack{E \\ y \in E}} f_x(E) \leqq 1 + (|E_0| - 1) = s$$

if $y \in E_0$. On the other hand,

$$\sum_E f(E) = 1 + \sum_{x \in E_0} \sum_E f_x(E) = 1 + |E_0| \cdot \nu(H) = 1 + s \cdot \nu(H),$$

which proves the assertion.

Now Theorem 5 easily follows from Theorem 7, 8 and (iii), (v). For assume H is not normal. Then we find a hypercritical partial hypergraph H' of it. For H' we have, obviously,

$$\tau(H') = \nu(H') + 1,$$

moreover, since H' is τ-critical,

$$\tau^*(H') < \tau(H')$$

and since H' is ν-critical,

$$\tau^*(H') > \nu(H') = \tau(H') - 1,$$

showing that $\tau^*(H')$ is not an integer.

Q.E.D.

There is another consequence of Theorems 7, 8 to be mentioned.

Theorem 9. A hypercritical hypergraph H has all its edges of size $\geq \delta(H)$.

Proof: Let H be hypercritical, then

$$\tau^*(H) \leq \tau(H) - 1 + \frac{1}{\delta(H)}$$

by Theorem 7; on the other hand

$$\tau^*(H) \geq \nu(H) + \frac{1}{|E_0|} = \tau(H) - 1 + \frac{1}{|E_0|}$$

for any $E_0 \in H$ by Theorem 8.

Q.E.D.

This theorem is interesting because the main conjecture (A) is equivalent to the following statement: The hypercritical hypergraphs are the following: (a) all (n-1)-tuples of an n-tuple, (b) odd circuits C_{2n+1}, (c) $C(\overline{C}_{2n+1})$, and (d) all hypergraphs arising from these by repeated application of the following construction: Take a new vertex, consider a family of edges with non-empty intersection, and add this new vertex to these edges.

III. Proofs of the Main Theorem.

Now we give the proof of the two main Theorems 1 and 2. We need a Lemma.

Lemma 1. Multiplication of edges of a q-normal hypergraph H yields q-normal hypergraphs.

Proof: It is enough to show this if only one new edge E', parallel to an edge E of H, is added to H. Let H_1 be the obtained hypergraph. It is enough to show that $q(H_1^*) = \delta(H_1^*)$, since for the partial hypergraphs of H, this follows similarly.

Obviously,

$$q(H) = \delta(H) \leqq \delta(H_1) \leqq q(H_1) \leqq q(H) + 1,$$

thus the only case we have to deal with is

$$\delta(H_1) = \delta(H) \qquad q(H_1) = q(H) + 1.$$

Consider a $q(H)$-coloring of H. Let $E_1, E_2, \ldots, E_k$ be the edges of the same color as E $(E_i \neq E)$. Let $H_2 = H - \{E_1, E_2, \ldots, E_k\}$. Then $\delta(H_2) = \delta(H) - 1$; for, consider a vertex x of maximum valency in H. Then x is contained in an edge of each color, in particular x belongs to one of the edges $E, E_1, \ldots, E_k$. However, $\delta(H_1) = \delta(H)$ implies that $x \notin E$.

Consider now a $(\delta(H) - 1)$-coloring of H_2, and add the edges $E', E_1, \ldots, E_k$ as a further color. This way we obtain a $\delta(H)$-coloring of H_1, a contradiction.

Q.E.D.

I. To prove Theorem 1, first we show that q-normal hypergraphs are normal. Obviously, it is enough to show they are semi-normal (since this holds then for the partial hypergraphs as well), thus by Theorem 6 we have to show $\nu(H) = \nu^*(H)$ (since this follows for its section hypergraphs similarly), i.e., $\nu_s(H) = s \cdot \nu(H)$ for any s. Consider an s-matching $f(E)$, and multiply every edge E of H by $f(E)$. The obtained hypergraph H_1 has

$$q(H_1) = \delta(H_1) \leqq s$$

since it is q-normal by Lemma 1. Moreover,

$$\nu(H_1) \leqq \nu(H).$$

Hence

$$\nu_s(H) = |H_1| \leqq \delta(H_1) \cdot \nu(H_1) \leqq s \cdot \nu(H).$$

Q.E.D.

The converse part of the theorem is, in fact, equivalent to the one proved above. For assume H is normal. Then, by Proposition 1, $\overline{L(H)}$ is perfect, hence $C(L(H))$ is q-normal. But then $C(L(H))$ is normal, hence $L(H)$ is perfect, i.e., H is q-normal.

Remark: Lemma 1 can be formulated as follows: Every s-matching of a normal hypergraph is the sum of s 1-matchings.

II. The proof of Theorem 2 is somewhat more complicated. We again have to prove $\nu_s(H) = s \cdot \nu(H)$ only. We use induction on s; we may assume H is hypercritical. Consider an s-matching $f(E)$. If $f(E) \leqq 1$ for every E then, denoting by H' the partial hypergraph of edges with $f(E) = 1$,

$$\sum_E f(E) \leqq |H'| \leqq \delta(H') \cdot \nu(H') \leqq s \cdot \nu(H),$$

by the assumption. Assume now there is an edge E_0 with $f(E_0) = k > 1$. Let $H_1 = H - \{E_0\}$. Then f gives an s-matching of H_1. Since H_1 is normal, f is the sum of s 1-matchings:

$$f(E) = \sum_{i=1}^{s} f_i(E) \qquad (E \neq E_0)$$

Assume the indices are chosen so that $\sum_{E \neq E_0} f_1(E) \geqq \ldots \geqq \sum_{E \neq E_0} f_s(E)$.

Consider now

$$f_0(E) = \begin{cases} \sum_{i=1}^{s-1} f_i(E) & \text{if } E \neq E_0, \\ k-1 & \text{if } E = E_0 . \end{cases}$$

Now $f_0(E)$ is an $(s-1)$-matching; really,

$$\sum_{x \in E} f_0(E) = \sum_{i=1}^{s-1} \sum_{x \in E} f_i(E) \leqq s-1 \text{ if } x \notin E_0,$$

$$\sum_{x \in E} f_0(E) \leqq k-1 + \sum_{\substack{x \in E \\ E \neq E_0}} f(E) = k-1 + \sum_{x \in E} f(E) - f(E_0) \leqq s-1 \text{ if } x \in E_0.$$

Thus

$$\sum_E f_0(E) \leqq (s-1)\nu(H),$$

$$\sum_{E \neq E_0} \sum_{i=1}^{s-1} f_i(E) \leqq (s-1)\nu(H) - (k-1)$$

whence it follows that $\sum_E f_x(E) < \nu(H)$, thus

$$\sum_E f(E) = f(E_0) + \sum_{i=1}^{s} \sum_{E \neq E_0} f_i(E) = k + \sum_{E \neq E_0} f_s(E) + \sum_{E \neq E_0} \sum_{i=1}^{s-1} f_i(E)$$

$$\leqq k + \nu(H) - 1 + (s-1)\nu(H) - (k-1) = s \cdot \nu(H), \qquad \text{Q.E.D.}$$

IV. Concluding remarks. A. All theorems except the 3rd and the 6th are of the form "if something holds for each partial hypergraph then something else holds for each partial hypergraph", Theorem 6 is of similar form with section hypergraph. It seems that this is the general form of theorems here; in other words, the interesting classes of hypergraphs contain, with a hypergraph H, some "smaller" hypergraphs. The first class of this type, investigated by Berge [2], is the class of hypergraphs without odd circuits; this class is closed under the very general operation of taking a subset of all edges. The next, more interesting class is the class of all balanced hypergraphs, introduced by Berge [2], and investigated also by Berge and Las Vergnas [4]; this class contains all partial subhypergraphs of a hypergraph of it. The class of normal hypergraphs contains all the partial hypergraphs of an element of it. The next step would be, probably, the class of all seminormal hypergraphs, closed for section hypergraphs only. Considering the increasing difficulty of proving useful theorems for these classes, this class of hypergraphs can lead to very deep questions.

B. Instead of the pair τ, ν one could consider other pairs related by max-min inequality; e.g., the stability number α and point covering number ρ; and investigate the class of all hypergraphs H such that $\alpha(H') = \rho(H')$ holds for each partial hypergraph H'.

C. The notions of balanced, normal etc., hypergraphs generalize the notion of bipartite graphs. There are many examples showing that for the most natural generalization, the class of 2-chromatic hypergraphs, no reasonable minimax theorem can be proved.

In the converse direction, balanced and normal hypergraphs are 2-chromatic,[5] as shown in Berge [2], and Fournier and Las Vergnas [6], respectively. Since any hypergraph in which the one-element sets are edges is seminormal, this is not true in this case any more. However, seminormal hypergraphs without one-element edges are, probably, 2-chromatic. This is true for the two special classes mentioned in the introduction.

[5] When speaking about colorings of hypergraphs, one-element edges are simply ignored.

REFERENCES

1. Berge, C., Färbung von Graphen deren sämtliche bzw. ungerade Kreise starr sind (Zusammenfassung) Wiss. Z. Martin-Luther-Univ. Halle-Wittenberg, Math. - Natur. Reihe (1961), 114-115.

2. Berge, C., Sur certains hypergraphes generalisant les graphes bipartites, Comb. Theory and its Appl., Bolyai J. Math. Soc. - North Holland Publ. Co. Budapest-Amsterdam-London, 1970, 119-133.

3. Berge, C., Graphes et hypergraphes, Dunod, Paris, 1970.

4. Berge, C. and Las Vergnas, M., Sur un theoreme du type König pour hypergraphes, Int. Conf. on Comb. Math., New York, 1970, 32-40.

5. Edmonds, J., Oral Communication at the Hypergraph Seminar, Columbus, September 1972.

6. Fournier, J. C., Las Vergnas, M., Une classe d'hypergraphes bichromatiques, Discret Math. 2(1972) 407-410.

7. Fulkerson, D. R. Anti-blocking pairs of polyhedra, to appear in J. Comb. Theory.

8. Gallai, T., Graphen mit triangularbaren ungeraden Vielecken, Mat. Kut. Int. Közl. 7(1962) 3-36.

9. Gallai, T., Kritische Graphen II, Mat. Kut Int. Közl. 8 (1963) 373-395.

10. Lovász, L. Normal hypergraphs and the perfect graph conjecture, Discrete Math 2(1972) 253-267.

11. Lovász, L., A characterization of perfect graph, to appear in J. Comb. Theory.

12. Sachs, H., On the Berge conjecture concerning perfect graphs, Comb. Structures and their Appl., Gordon and Breach, New York-London-Paris, 1969, 377-384.

QUELQUES PROBLEMES CONCERNANT LES CLIQUES DES HYPERGRAPHES h-COMPLETS ET q-PARTI h-COMPLETS

J.C. Meyer, Université de Paris-Sud Centre d'Orsay

1. Introduction. Soient n et h deux entiers, avec $h \leq n$. L'hypergraphe simple ayant pour ensemble de sommets un ensemble X à n éléments et pour ensemble d'arêtes l'ensemble $\mathcal{P}_h(X)$ des h-parties de X est noté K_n^h et s'appelle l'hypergraphe h-complet à n sommets.

Soient q et h deux entiers, avec $h \leq q$. L'hypergraphe simple ayant pour ensemble de sommets la réunion d'ensembles disjoints $X_1, X_2, \ldots X_q$ et pour arêtes tous les sous-ensembles E de $\bigcup_{i=1}^{q} X_i$ de cardinalité h tels que $|E \cap X_i| \leq 1$ pour tout i, $1 \leq i \leq q$, s'appelle l'hypergraphe q-parti h-complet sur $\{X_1, X_2, \ldots, X_q\}$ et est noté $K^h_{n_1, n_2, \ldots, n_q}$, où $n_i = |X_i|$ pour tout i, $1 \leq i \leq q$. On supposera dans la suite $n_1 \leq n_2 \leq \ldots \leq n_q$. Lorsque $n_i = n$ pour tout i, $1 \leq i \leq q$, on notera cet hypergraphe $K^h_{q \times n}$.

Remarques.

. Lorsque $q = h$, $K^h_{n_1, n_2, \ldots, n_q}$ s'appelle l'hypergraphe h-parti complet sur $\{X_1, X_2, \ldots, X_q\}$

. $K^h_{n \times 1} = K^h_n$

. Lorsque $h=2$, K^2_n est le graphe complet K_n.

. Lorsque $h=q=2$, $K^2_{n_1, n_2}$ est le graphe biparti complet K_{n_1, n_2}.

. Si $q' > q$, un hypergraphe q-parti h-complet sur $\{X_1, X_2, \ldots, X_q\}$ peut être considéré comme un hypergraphe q'-parti h-complet sur $\{Y_1, \ldots, Y_{q'-q}, X_1, \ldots, X_q\}$ où $Y_1 = \ldots = Y_{q'-q} = \emptyset$.

Etant donné un hypergraphe $H = (E_i \mid i \in I)$ on appelle clique de H un hypergraphe partiel $\mathcal{C} = (E_j \mid j \in J)$ de H tel que $E_k \cap E_\ell = \emptyset$ pour k et ℓ dans J.

Soit $\mathcal{C} = (E_j | j \in J)$ une clique maximale d'un hypergraphe $H = (E_i | i \in I)$. Si $\bigcap_{j \in J} E_j = \emptyset$, nous dirons que la clique $\mathcal{C}$ est <u>secondaire</u>. Si

$\bigcap_{j \in J} E_j = \{a\}$, nous dirons que la clique $\mathcal{C}$ est <u>associée au sommet</u> a.

On note par

$\omega(H)$ <u>le plus grand nombre d'arêtes d'une clique de</u> H

$\omega'(H)$ <u>le plus petit nombre d'arêtes d'une clique maximale de</u> H

$t(H)$ <u>le plus petit entier</u> t <u>pour lequel il existe une partition</u> $(I_1, I_2, \ldots, I_t)$ <u>de</u> I <u>tel que les hypergraphes partiels de</u> H $(E_i | i \in I_\alpha)$, <u>pour</u> $1 \leq \alpha \leq t$, <u>soient des cliques de</u> H. $t(H)$ est aussi le plus petit entier t pour lequel il existe un recouvrement $(I_1, I_2, \ldots, I_t)$ de I tel que les hypergraphes partiels de H, $(E_i | i \in I_\alpha)$, pour $1 \leq \alpha \leq t$, soient des cliques de H. Avec les notations de [1] on a $t(H) = \Theta(L(H))$.

Si α est un nombre réel positif, $[\alpha]^*$ désigne <u>la partie entière par excès de</u> α.

Dans cet article on se propose de rappeler et de donner quelques propriétés des nombres $\omega(H)$, $\omega'(H)$, $t(H)$, lorsque H est un hypergraphe h-complet, q parti h-complet et h-parti complet.

2. <u>Hypergraphe</u> K_n^h.

P. Erdös, Chao Ko et R. Rado [2] ont montré que

$$\omega(K_n^h) = \begin{cases} \binom{n}{h} & \text{si } n < 2h \\ \binom{n-1}{h-1} & \text{si } n \geq 2h. \end{cases}$$

<u>Proposition 1.</u> <u>Toutes les cliques maximales de</u> K_{2h}^h <u>ont</u> $\binom{2h-1}{h}$ <u>arêtes</u>.

En effet soit $\mathcal{C} = (E_j | j \in J)$ une clique maximale de K_{2h}^h. $\mathcal{C}$ étant une clique de K_{2h}^h , $\overline{\mathcal{C}} = (X - E_j | j \in J)$ est une clique de K_{2h}^h. $\mathcal{C}$ étant une clique maximale de K_{2h}^h il est facile de montrer que $\mathcal{C} \cup \overline{\mathcal{C}} = \mathcal{P}_h(X)$. Comme $|\mathcal{C}| = |\overline{\mathcal{C}}|$ on otbient $|\overline{\mathcal{C}}| = \frac{1}{2}\binom{2h}{h} = \binom{2h-1}{h}$. C.Q.F.D.

<u>Proposition 2.</u> <u>Si</u> $h = p^\alpha + 1$, <u>où</u> p <u>est un nombre premier et</u> α <u>un entier et</u>

<u>et si</u> $n \geq (h-1)^2+h$, <u>alors</u> $\omega'(K_n^h) \leq (h-1)^2 + h$.

En effet soit P_h un plan projectif d'ordre h-1. On peut considérer P_h comme un hypergraphe d'ordre $(h-1)^2+h$, ses sommets étant les points du plan projectif, ses arêtes les droites du plan projectif.

Puisque $n \geqslant (h-1)^2+h$, on peut supposer que P_h est un hypergraphe partiel de K_n^h. P_h est ainsi une clique de K_n^h ayant $(h-1)^2+h$ arêtes. Montrons qu'elle est maximale. Soit E une arête de K_n^h qui ne soit pas une arête de P_h et montrons qu'il existe une arête de P_h qui ne rencontre pas E. Si E a au plus un sommet dans P_h ceci est bien clair. Dans le cas contraire posons $E = \{a_1, a_2, \ldots, a_h\}$ et supposons que a_1 et a_2 soient des sommets de P_h. Par les points a_1 et a_2 il passe une arête F_o de P_h. Puisque E n'est pas une arête de P_h soit $b \in F_o - E$. Par le point b il passe h-1 arêtes $F_1, F_2, \ldots, F_{h-1}$ de P_h autres que F_o. Les ensembles $F_i - \{b\}$, $0 \leq i \leq h-1$ étant deux à deux disjoints, on a

$$h = |E| \geq \sum_{i=o}^{h-1} |F_i \cap E| \geq 2 + \sum_{i=1}^{h-1} |F_i \cap E|.$$

Donc

$$\sum_{i=1}^{h-1} |F_i \cap E| \leq h-2,$$

et par suite il existe un i_o, $I \leq i_o \leq h-1$, tel que $E \cap F_{i_o} = \emptyset$ C.Q.F.D.

<u>Théorème 1</u>. <u>Si</u> $n \geq 7$, $\omega'(K_n^3) = 7$.

Soit $\mathcal{C} = (E_j | j \in J)$ une clique maximale de K_n^3. D'après la proposition 2, il nous suffit de montrer que $|\mathcal{C}| \geq 7$. Nous allons montrer que l'hypothèse $|\mathcal{C}| \leq 6$ est absurde.

Posons $Y = \bigcup_{j \in J} E_j$. On a $|Y| \geq 6$ car si $|Y| < 6$ soit Y' tel que $Y \subset Y'$ et $|Y'| = 5$. Alors $\mathcal{P}_3(Y')$ est une clique de K_n^3 contenant $\mathcal{C}$ et de cardinalité $\binom{5}{3} = 10$. On a $|Y| \geq 7$ car si $|Y| = 6$ on a d'après la proposition 1, $|\mathcal{C}| = \binom{5}{3} = 10$. On a donc $|Y| \geq 7$.

Si $x \in Y$, appelons <u>degré de x</u> (dans $\mathcal{C}$) et notons par d(x) le nombre d'arêtes de $\mathcal{C}$ contenant x.

1) <u>Tous les points de</u> Y <u>sont de degré</u> ≥ 2.

En effet soit x un point de degré 1 et soit E l'arête de $\mathcal{C}$ contenant x. Posons $E-\{x\} = \{a,b\}$. $\{a,b\}$ rencontre toutes les arêtes de $\mathcal{C}$ et les parties $\{a,b,y\}$ où $y \in X-\{a,b\}$ sont donc des arêtes de $\mathcal{C}$ Il y en a au moins $|X|-|E-\{x\}| \geq 5$. On peut supposer que ni a, ni b appartiennent à toutes les arêtes de $\mathcal{C}$ car sinon $|\mathcal{C}| \geq \binom{6}{2} = 15$. Donc il y a une arête de $\mathcal{C}$ qui ne contient pas a (et qui contient alors b) et une arête de $\mathcal{C}$ qui ne contient pas b (et qui contient alors a). Donc $|\mathcal{C}| \geq 7$.

2) <u>Tous les points de</u> Y <u>sont de degré</u> ≥ 3.

En effet soit x un point de degré 2 et soient E_1 et E_2 les deux arêtes de $\mathcal{C}$ contenant x. On a $|E_1 \cap E_2| = 1$ car sinon $E_1-\{x\}$ rencontre toutes les arêtes de $\mathcal{C}$ et on aurait comme précédemment $|\mathcal{C}| \geq 7$. Posons $E_1 = \{x,a,b\}$, $E_2 = \{x,\alpha,\beta\}$ $\{a,b\}$ rencontre toutes les arêtes de $\mathcal{C}-\{E_2\}$. Donc $F_1 = \{a,b,\alpha\}$, $F_2 = \{a,b,\beta\}$ sont des arêtes de $\mathcal{C}$. De même $F_3 = \{\alpha,\beta,a\}$ et $F_4 = \{\alpha,\beta,b\}$ sont des arêtes de $\mathcal{C}$. Puisque $|Y| \geq 7 > |\{x,a,b\ \alpha,\beta\}|$ il y a encore au moins une autre arête dans et $|\mathcal{C}| \geq 7$.

Tous les points de Y étant de degré ≥ 3 et $|Y| \geq 7$, on a $|\mathcal{C}| \geq 7$. C.Q.F.D.

<u>Conjecture 1. Soit</u> $h \geq 4$ <u>tel que</u> $h = p^\alpha + 1$, <u>où</u> p <u>est un nombre premier et</u> α <u>un entier, et soit</u> $n \geq (h-1)^2+h$. <u>Alors</u> $\omega'(K_n^h) = (h-1)^2 + h$.

<u>Proposition 3. On a</u>

$$t(K_n^h) = 1 \quad \text{si} \quad h \leq n \leq 2h-1$$

$$t(K_n^h) = 2 \quad \text{si} \quad n = 2h$$

$$t(K_n^h) \leq n-2h+2 \quad \text{si} \quad n \geq 2h.$$

En effet : si $h \leq n \leq 2h-1$, K_n^h est une clique .

si $n = 2h$, il existe deux arêtes disjointes dans K_n^h, donc $t(K_n^h) \geq 2$. En prenant la clique associée à un sommet a et la clique $(X-\{a\}, \mathcal{P}_h(X-\{a\}))$ on obtient $t(K_n^h) = 2$.

Soit $n > 2h$ et supposons que $t(K_{n-1}^h) \leq (n-1) - 2h+2$.

Soit $\mathcal{C}_o$ la clique associée à un sommet $a \in X$.

$(X-\{a\}, \mathcal{P}_h(X-\{a\})) \simeq K^h_{n-1}$ et il existe d'après l'hypothèse de récurrence une famille $\mathcal{C}_1,..,\mathcal{C}_k$ de cliques avec $k \leq (n-1)-2h+2$ telle que

$$\bigcup_{i=1}^{k} \mathcal{C}_i = \mathcal{P}_h(X-\{a\}).$$

On a alors $\bigcup_{i=o}^{k} \mathcal{C}_i = \mathcal{P}_h(X)$ et $t(K^h_n) \leq n-2h+2$. C.Q.F.D.

<u>Proposition 4</u>. <u>Si</u> $n \geq 3$, <u>on a</u> $t(K^2_n) = n-2$.

Nous allons démontrer ce résultat par récurrence sur n. Si $\mathcal{C}$ est une clique secondaire on a $|\mathcal{C}| = 3$. Supposons le résultat vrai pour n-1, avec $n > 4$. Soit $\mathcal{C}_1,...,\mathcal{C}_k$ un recouvrement de $\mathcal{P}_2(X)$ par des cliques que l'on peut supposer maximales et supposons $k < n-2$. Si $n > 4$, on vérifie que $[\frac{1}{3}\binom{n}{2}]^* > n-3$ et par suite il existe au moins une clique associée à un sommet a, par exemple $\mathcal{C}_1$. Mais alors $\bigcup_{i=2}^{k} \mathcal{C}_i = \mathcal{P}_2(X-\{a\})$ et $t(K^2_{n-1}) < (n-1)-2$ ce qui est absurde. C.Q.F.D.

<u>Propositinn 5</u>. <u>Si</u> $n \geq 5$, <u>on a</u> $t(K^3_n) = n-4$.

Nous allons démontrer ce résultat par récurrence sur n. Si $\mathcal{C}$ est une clique secondaire, on a d'après un résultat de A.J.W. Hilton et E.C. Milner [3] : $|\mathcal{C}| \leq 3n-8$. Supposons le résultat vrai pour n-1, avec $n > 6$. Soit $\mathcal{C}_1,...,\mathcal{C}_k$ un recouvrement de $\mathcal{P}_3(X)$ par des cliques maximales et supposons $k < n-4$. Si $n > 6$, on véfifie que $\left[\frac{1}{3n-8} \binom{n}{3}\right]^* > n-5$ et par suite il existe au moins une clique associée à un sommet a, par exemple $\mathcal{C}_1$. Mais alors $\bigcup_{i=2}^{k} \mathcal{C}_i = \mathcal{P}_3(X-\{a\})$ et $t(K^3_{n-1}) < (n-1)-4$ ce qui est absurde.
C.Q.F.D.

<u>Conjecture</u>. (M. Kneser [4]).

Si $h \geq 4$ et $n > 2h$, $t(K^h_n) = n-2h+2$.

3. <u>Hypergraphe</u> $K^h_{n_1,n_2,...,n_q}$.

<u>Classification des cliques maximales de</u> $K^h_{n_1,n_2,...n_h}$.

Si $n_1 = 1$, $K^h_{n_1,n_2,...,n_h}$ est une clique.

Si $n_1 > 1$, les cliques maximales sont de différents types et il convient de les classifier.

Soit $R \subset R_o = \{1,2,\ldots,h\}$ avec $R \neq R_o$ et $R \neq \emptyset$. On dira qu'un ensemble d'arêtes $\mathcal{F}$ de $K^h_{n_1,n_2,\ldots,n_h}$ est du type $\mathcal{C}_R$ si $\mathcal{F}$ est formé d'une part par toutes les arêtes qui contiennent un ensemble fixé $A_o = \{a_i | i \in R\}$ avec $a_i \in X_i$ pour $i \in R$ et d'autre part par toutes les arêtes de la forme $A \cup B$ avec $A \cap A_o \neq \emptyset$ et $B \in \mathcal{B}_o$, où $\mathcal{B}_o$ est une clique maximale fixée de l'hypergraphe (h-r)-parti complet sur $\{X_i | i \in R_o - R\}$ $(r = |R|)$. Une clique maximale pour laquelle on ne peut définir un tel A_o sera dite du type $\mathcal{C}_\emptyset$

Proposition 6. Soit $R \subset R_o$, $R \neq R_o$. Si $n_1 > 1$, toute famille d'arêtes $\mathcal{F}$ du type $\mathcal{C}_R$ est une clique maximale. De plus toute clique maximale est du type $\mathcal{C}_R$ pour au moins un $R \subset R_o$, $R \neq R_o$.

On peut supposer $R \neq \emptyset$ car si $\mathcal{F} \in \mathcal{C}_\emptyset$, $\mathcal{F}$ est une clique maximale par définition.

$\mathcal{F}$ est une clique car toutes les arêtes de $\mathcal{F}$ se rencontrent deux à deux. Pour montrer que $\mathcal{F}$ est maximale soit E_1 une arête n'appartenant pas à $\mathcal{F}$ et montrons qu'il existe une arête $F_1 \in \mathcal{F}$ telle que $E_1 \cap F_1 = \emptyset$. Posons $E_1 = A_1 \cup B_1$ où $A_1 \subset \bigcup_{i \in R} X_i$ et $B_1 \subset \bigcup_{i \in R_o - R} X_i$.

Si $A_1 \cap A_o = \emptyset$, prenons $B \subset X_{R_o - R}$ de sorte que $B \cap B_1 = \emptyset$ (ceci est possible car $n_1 > 1$). Alors l'arête $A_o \cup B$ appartient à $\mathcal{F}$ et est disjointe de E_1.

Si $A \cap A_o \neq \emptyset$, alors $B_1 \notin \mathcal{B}_o$ et il existe $B \in \mathcal{B}_o$ tel que $B \cap B_1 = \emptyset$. D'autre part pour $i \in R$, soit $x_i \in X_i$ tel que :

$$x_i = a_i \quad \text{si} \quad a_i \notin A_1$$
$$x_i \neq a_i \quad \text{si} \quad a_i \in A_1$$

(on rappelle ici que $A_o = \{a_i | i \in R\}$).

L'ensemble $A_2 = \{x_i | i \in P\}$ rencontre A_o (car $A_1 \neq A_o$ puisque $E_1 \notin \mathcal{F}$) et est disjoint de A_1. Donc l'arête $F_1 = A_2 \cup B$ appartient à $\mathcal{F}$ et est disjointe de E_1.

C.Q.F.D.

Proposition 7. Toutes les cliques maximales de $K^h_{h \times 2}$ ont 2^{h-1} arêtes.

La démonstration est analogue à celle de la proposition 1.

Théorème 2. Si $n \geq 2$, $\omega(K^h_{q \times n}) = \binom{q-1}{h-1} n^{h-1}$.

Tout d'abord une clique associée à un sommet a une cardinalité égale à $\binom{q-1}{h-1} n^{h-1}$. Il nous suffit donc de montrer que $\omega(K^h_{q \times n}) \leq \binom{q-1}{h-1} n^{h-1}$.

Si $q=h$, ceci résulte d'un résultat de C. Berge [5].

Si $h=2$ et $q > 2$, ceci résulte du fait que les seules cliques secondaires sont les triangles.

Soit (h,q) avec $q > h$. Supposons le résultat démontré pour les couples (h',q'), $h' \leq q'$ où $(h',q') < (h,q)$ ($(h',q') < (h,q)$ si $h' \leq h$ et $q' \leq q$ l'une au moins des deux inégalités étant strictes).

Soit $\mathcal{C} = (E_i \mid I \leq i \leq p)$ une clique de $K^h_{q \times n}$. p étant fixé choisissons $\mathcal{C}$ telle que $\sum_{i=1}^{p} |E_i \cap X_q|$ soit minimum.

Cas 1. Supposons que si $E \cap X_q \neq \emptyset$, $E \in \mathcal{C}$ et $x \in \bigcup_{k=1}^{q-1} X_k - E$, alors $E - X_q + \{x\} \in \mathcal{C}$.

On peut supposer que

$$E_i \cap X_q \neq \emptyset \quad \text{pour } I \leq i \leq p_o$$

et

$$E_i \cap X_q = \emptyset \quad \text{pour } p_o < i \leq p.$$

Posons $E'_i = E_i - X_q$ si $I \leq i \leq p_o$. Si $I \leq i < j \leq p_o$, $E'_i \cap E'_j \neq \emptyset$. En effet soit X_k tel que $E_i \cap X_k = \emptyset$ et soit $x \in X_k$ tel que $x \notin E_j$ (ceci est possible car $|X_k| \geq 2$). Alors

$(E_i - X_q + \{x\}) \cap E_j \neq \emptyset$ car $E_i - X_q + \{x\} \in \mathcal{C}$ par hypothèse. Or puisque $x \notin E_j$, $(E_i - X_q) \cap E_j \neq \emptyset$ et donc $E'_i \cap E'_j \neq \emptyset$.

Considérons les E'_i distincts et appelons-les $F_1, \ldots, F_{p_1}$. On a $p_o \leq n p_1$ $(F_1, \ldots, F_p)$ est une clique de $K^{h-1}_{(q-1)\times n}$. Donc d'après l'hypothèse de récurrence on a

$$p_1 \leq \binom{q-2}{h-2} n^{h-2}.$$

Donc

$$p_o \leq \binom{q-2}{h-2} n^{h-1}.$$

D'autre part $(E_{p_o+1}, \ldots, E_p)$ est une clique de $K^h_{(q-1)\times n}$. Donc d'après l'hypothèse de récurrence on a :

$$p - p_o \leq \binom{q-2}{h-1} n^{h-1}.$$

Par suite

$$p \leq \binom{q-2}{h-2} n^{h-1} + \binom{q-2}{h-1} n^{h-1} = \binom{q-1}{h-1} n^{h-1}.$$

Cas 2. (opposé du cas 1). Supposons qu'il existe $E \in \mathcal{C}$ telle que $E \cap X_q = \{y\}$ et $x \in \bigcup_{k=1}^{q-1} X_k - E$ tel que $E - X_q + \{x\} \notin \mathcal{C}$. Nous pouvons supposer que

$y \in E_i$, $x \notin E_i$ $\quad E'_i = E_i - \{y\} + \{x\} \notin \mathcal{C}$ pour $I \leq i \leq p_1$

$y \in E_i$, $x \notin E_i$ $\quad E_i - \{y\} + \{x\} \in \mathcal{C}$ pour $p_1 < i \leq p_2$

$y \in E_i$, $x \in E_i$ pour $p_2 < i \leq p_3$

$y \notin E_i$ pour $p_3 < i \leq p$.

Posons $F_i = E'_i$ pour $I \leq i \leq p_1$ et $F_i = E_i$ pour $p_1 < i \leq p$. On vérifie que $(F_1,\ldots,F_p)$ est une clique de cardinalité p telle que

$$\sum_{i=1}^{p} |F_i \cap X_q| < \sum_{i=1}^{p} |E_i \cap X_q|.$$

Donc le cas 2 n'est pas possible. C.Q.F.D.

Remarque. Cette démonstration est inspirée de la démonstration du théorème 1 dans l'article de P. Erdös, Chao Ko, R. Rado [2].

Proposition 8. Si $h = p^\alpha + 1$, où p est un nombre premier et α un entier et si $n_{q-h+1} \geq h-1$, alors $\omega'(K^h_{n_1,n_2,\ldots,n_q}) \leq (h-1)^2$.

Considérons un plan projectif P_h d'ordre $h-1$. Soit $a \in P_h$ et soient $D_1,\ldots,D_h$ les droites de P_h passant par a. Posons : $Y_1 = D_1 - \{a\}, \ldots, Y_h = D_h - \{a\}$.

Considérons l'hypergraphe $\mathcal{C} = (Y, \mathcal{E})$ où $Y = \bigcup_{i=1}^{h} Y_i$ et $E \in \mathcal{E}$ si et seulement si E est une droite de P_h contenue dans Y.

Si $E \in \mathcal{E}$, $|E \cap Y_i| = 1$ pour tout i, $I \leq i \leq h$, car $E \cap D_i \neq \emptyset$ et $a \notin E$.

Puisque $n_{q-h+1} \geq h-1$, on peut supposer $Y_1 \subset X_{q-h+1}, \ldots, Y_h \subset X_q$. $\mathcal{C}$ est ainsi un hypergraphe partiel de $K^h_{n_1,n_2,\ldots,n_q}$. Nous allons montrer que $\mathcal{C}$ est une clique maximale de cardinalité $(h-1)^2$. $\mathcal{C}$ est bien une clique car

deux droites du plan projectif ont toujours un point commun et d'autre part si $E \in \mathcal{E}$ et $E' \in \mathcal{E}$, $a \notin E \cap E'$.
On a $|\mathcal{C}| = (h-1)^2$.
Soit F une arête de $K^h_{n_1,n_2,\dots,n_q}$, $F \notin \mathcal{E}$.

Cas 1: $|F \cap Y| \leq 1$. Soit D une droite du plan projectif passant par a et $F \cap Y$. Soit b un point de D, $b \neq a$ et $b \notin F \cap Y$. Soit $E \in \mathcal{E}$ telle que $b \in E$. Alors $E \cap F = \emptyset$.

Cas 2: $|F \cap Y| \geq 2$. On peut supposer par exemple que $|F \cap Y_1| = |F \cap Y_2| = 1$. Soient $a_1 \in F \cap Y_1$, $a_2 \in F \cap Y_2$. Soit E l'arête de $\mathcal{C}$ passant par les points a_1 et a_2. Puisque $F \notin \mathcal{E}$, soit $b \in E$ tel que $b \notin F$. On peut supposer $b \in Y_3$.

Ou bien $|F \cap Y_3| = 1$ et dans ce cas soit $c \in F \cap Y_3$.

Ou bien $F \cap Y_3 = \emptyset$ et il existe $\ell \leq q-h$ tel que $|F \cap X_\ell| = 1$ et dans ce cas soit $c \in F \cap X_\ell$ (dans ce cas $c \notin Y$ car $Y \subset \bigcup_{j=q-h+1}^{q} X_j$).

Par le point b il passe $h-2$ arêtes $E_1,\dots,E_{h-2}$ de $\mathcal{C}$ autres que E. On a $\{a_1,a_2,c\} \cap (\bigcap_{i=1}^{h-2} E_i) = \emptyset$, et les ensembles $E_i-\{b\}$ sont pour $i=1,\dots,h-2$ deux à deux disjoints. Il existe donc un i_o, $1 \leq i_o \leq h-2$, tel que $E_{i_o} \cap F = \emptyset$.

C.Q.F.D.

Théorème 3. Si $n_{q-2} \geq 2$, $\omega'(K^3_{n_1,n_2,\dots,n_q}) = 4$.

D'après la proposition 8 il nous suffit de montrer que toute clique maximale de $K^3_{n_1,n_2,\dots,n_q}$ a au moins 4 arêtes. Soit $\mathcal{C}$ une clique $K^3_{n_1,n_2,\dots,n_q}$ et supposons $|\mathcal{C}| \leq 3$. Nous allons montrer que $\mathcal{C}$ n'est pas une clique maximale. Il est clair que l'on peut, supposer $|\mathcal{C}| = 3$. Posons : $\mathcal{C} = (E_1,E_2,E_3)$.

Cas 1: $E_1 \cap E_2 \cap E_3 \neq \emptyset$. Soit $a \in E_1 \cap E_2 \cap E_3$. Il y a dans $K^3_{n_1,n_2,\dots,n_q}$ au moins quatre arêtes passant par a et on peut compléter $\mathcal{C}$ en une clique ayant au moins quatre arêtes.

Cas 2: $E_1 \cap E_2 \cap E_3 = \emptyset$. Soit $a \in E_2 \cap E_3$, $b \in E_1 \cap E_3$ et $c \in E_1 \cap E_2$. a,b,c sont trois points distincts et ils sont dans des X_i différents. Alors $\mathcal{C}' = (E_1, E_2, E_3, \{a,b,c\})$ est une clique contenant $\mathcal{C}$ et ayant quatre arêtes.

C.Q.F.D.

Conjecture 2. Soit $h \geq 4$ tel que $h = p^\alpha + 1$, où p est un nombre premier et α un entier. Si $n_{q-h+1} \geq h-1$, alors $\omega'(K^h_{n_1,n_2,\ldots,n_q}) = (h-1)^2$.

Proposition 9. Si $h = p^\alpha + 1$ où p est un nombre premier et α un entier et si $n_1 \geq \max(3, h-1)$ les cliques maximales de cardinalité $\omega'(K^h_{n_1,n_2,\ldots,n_h})$ sont du type $\mathcal{C}_\emptyset$.

Soit $\mathcal{C}$ une clique de type $\mathcal{C}_R$ avec $R \neq \emptyset$, $R \subseteq R_o = \{1,2,\ldots,h\}$ définie par A_o et $\mathcal{B}_o$. Il nous suffit d'après la proposition 8 de montrer que $|\mathcal{C}| > (h-1)^2$.

Le nombre d'arêtes contenant A_o est égal à $\prod_{i \in R_o - R} n_i$ et d'autre part le nombre d'arêtes de la forme $A \cup B$ avec $A \cap A_o \neq \emptyset$, $A \neq A_o$ et $B \in \mathcal{B}_o$ est égal à

$$\left(\prod_{i \in R} n_i - \prod_{i \in R} (n_i - 1) - 1 \right) |\mathcal{B}_o| .$$

On a donc

$$|\mathcal{C}| \geq \prod_{i \in R_o - R} n_i + \left(\prod_{i \in R} n_i - \prod_{i \in R} (n_i - 1) - 1 \right) |\mathcal{B}_o|$$

et par suite

$$|\mathcal{C}| \geq \prod_{i \in R_o - R} n_i + \prod_{i \in R} n_i - \prod_{i \in R} (n_i - 1) - 1$$

Pour chaque $i \in R_o$, le second membre est une fonction croissante de n_i. Donc puisque $n_i \geq 3$ pour tout $i \subset R_o$, on a :

$$|\mathcal{C}| \geqslant 3^{h-r} + 3^r - 2^r - 1 \text{ , où } r = |R| .$$

Montrons que $\varphi(h) = 3^{h-r} + 3^r - 2^r - 1 - (h-1)^2 > 0$ pour $h \geq r+1$.

Nous avons $\varphi'(h) = (\text{Log } 3)\, 3^{h-r} - 2(h-1)$

$\varphi''(h) = \text{Log } 3)^2\, 3^{h-r} - 2 \geq 3(\text{Log}3)^2 - 2 > 0$ pour $h \geq r+1$.

Ainsi φ' est croissante.

<u>Cas</u> r=1. On a $\varphi'(2) > 0$, donc φ est croissante et $\varphi(2) = 2 > 0$. Donc $\varphi(h) > 0$ pour $h \geq 2$.

<u>Cas</u> $r \geq 2$. φ' s'annule pour $h_o \geq 3$ vérifiant :

$$(\text{Log } 3)3^{h_o-r} = 2(h_o-1)$$

et pour h_o on a $\varphi(h_o) = \dfrac{2(h_o-1)}{\text{Log } 3} + 3^r - 2^r -1-(h_o-1)^2$.

Nous allons montrer que $\varphi(h_o) > 0$. Posons

$$g(h) = (h-1)^2 - \frac{2(h-1)}{\text{Log } 3} = h^2 - 2\alpha h + \beta \quad \text{où} \quad \alpha = 1 + \frac{1}{\text{Log } 3} \text{ et } \beta = 1 + \frac{2}{\text{Log } 3}.$$

On a, $1,90 < \alpha < 1,91$ et $2,81 < \beta < 2,82$. Donc $\alpha^2 - \beta > 0$ et la plus grande racine de g est inférieure à 3. Donc dans l'intervalle $[3,+\infty[$ g est croissante.

On a $\varphi'(2r) = (\text{Log } 3)\, 3^r-2(2r-1) > 0$ pour $r \geq 2$. φ' étant croissante on a donc $h_o < 2r$. g étant croissante sur $[3,+\infty[$ on a $g(h_o) < g(2r)$. Nous allons montrer que $3^r-2^r-1-g(2r) > 0$ ce qui entraînera $\varphi(h_o) > 0$.

Soit $\lambda(r) = 3^r - 2^r - 1 - g(2r) = 3^r - 2^r - 1 -(4r^2-4\alpha r + \beta)$. On a $\lambda(2) > 0$. Supposons $r \geq 3$:

$$\lambda'(r) = (\text{Log } 3)\, 3^r - (\text{Log } 2)2^r - 8r + 4\alpha$$

$$\lambda''(r) = (\text{Log } 3)^2\, 3^r - (\text{Log } 2)^2\, 2^r-8.$$

λ'' est croissante et $\lambda''(3) > 0$. Donc λ' est croissante sur $[3,+\infty[$ et puisque $\lambda'(3) > 0$, λ est croissante. Puisque $\lambda(3) > 0$ on a bien $\varphi(h_o) > 0$.

C.Q.F.D.

<u>Proposition 10</u>. <u>Si</u> $n_q \geq 2$, <u>on a</u> $t(K^2_{n_1,n_2,\ldots,n_q}) = \sum_{i=1}^{q-1} n_i$.

En considérant les cliques associées aux sommets x pour $x \in \bigcup_{i=1}^{q-1} X_i$, on obtient $t(K^2_{n_1,n_2,\ldots,n_q}) \leq \sum_{i=1}^{q-1} n_i$. Nous dirons que $(n_1,n_2,\ldots,n_q) < (n'_1,n'_2,\ldots,n'_q)$ si $n_i \leq n'_i$ pour $I \leq i \leq q$ et $(n_1,n_2,\ldots,n_q) \neq (n'_1,n'_2,\ldots,n'_q)$. La proposition est vraie pour le graphe $K^2_{n_1,n_2,\ldots,n_q}$ avec $n_1 = \ldots = n_{q-2} = 0$, $n_{q-1} = 1$ et $n_q \geq 2$. Supposons-la vraie pour les q-uples inférieurs à $(n_1,n_2,\ldots,n_q)$ et montrons-la pour

$(n_1,n_2,\dots,n_q)$.

Supposons que $t(K^2_{n_1,n_2,\dots,n_q}) < \sum_{i=1}^{q-1} n_i$.

Montrons que dans un recouvrement en $t(K^2_{n_1,n_2,\dots,n_q})$ cliques maximales il y a au moins une clique associée à un sommet. S'il n'y a pas de clique associée à un sommet, toutes les cliques sont secondaires et donc de cardinalité 3. Ces cliques recouvrent donc au plus $3(\sum_{i=1}^{q-1} n_i - 1)$ arêtes. Or $K^2_{n_1,n_2,\dots,n_q}$ ayant $\sum_{i<j} n_i n_j$ arêtes on a :

$$\sum_{i<j} n_i n_j \leq 3(\sum_{i=1}^{q-1} n_i - 1) \quad (1)$$

Si $n_q \geq 3$, on a $\sum_{i<j} n_i n_j \geq \sum_{i<q-1} n_i n_q$ donc $\sum_{i<j} n_i n_j \geq 3 \sum_{i \leq q-1} n_i$ ce qui contredit (1).

Si $n_q = 2$, soit p le nombre de n_i non nuls et k le nombre de n_i égaux à 1. On a $0 \leq k \leq p-1$. Alors

$$\sum_{i<j} n_i n_j = \binom{k}{2} + 4\binom{p-k}{2} + 2k(p-k)$$

et

$$3(\sum_{i=1}^{q-1} n_i - 1) = 3k + 6(p-k-1) - 3.$$

Posons $P(k,p) = 2(\sum_{i<j} n_i n_j - 3(\sum_{i=1}^{q-1} n_i - 1))$ et montrons que $P(k,p) > 0$ pour $1 \leq k \leq p-1$ ce qui contredira à nouveau (1). On a

$$P(k,p) = k^2 + (9-4p)k + 4p^2 - 16p + 18$$

$$P'_k(k,p) = 2k + 9 - 4p = 0 \quad \text{si} \quad k = \frac{4p-9}{2}.$$

Si $\frac{4p-9}{2} \geqslant p-1$, soit $p \geq 4$, $P(k,p)$ est une fonction décroissante de k pour $0 \leq k \leq p-1$ et $P(k,p) \geq P(p-1,p) = p^2 - 5p + 10 > 0$. Si $p=2$, $P(k,2) = k^2 + k + 2 > 0$ et si $p=3$, $P(k,3) = k^2 - 3k + 6 > 0$ pour $k=0,1,2$. Donc il y a au moins une clique associée à un sommet $a \in X_{i_0}$. On a alors $t(K^2_{n_1,n_2,\dots,n_q}) = t(H) + 1$, où H est le graphe q-parti complet sur

$\{X_1,\ldots,X_{i_o}-\{a\},\ldots,X_q\}$ et alors $t(H) < \sum_{i=1}^{q-1} n_i - 1$, ce qui contredit la proposition 4 si $i_o = q$, $n_q=2$ et tous les autres n_i non nuls égaux à 1, et l'hypothèse de récurrence dans les autres cas.

C.Q.F.D.

REFERENCES

1. C. Berge, Graphes et Hypergraphes, Dunod, Paris, 1970.

2. P. Erdös, Chao Ko, R. Rado, Intersection theorems for systems of finite sets, Quart. J. of Math. (Oxford) (2) 12 (1961) 313-320.

3. A.J.W. Hilton and E.C. Milner, Some Intersection theorems for systems of finite sets, Quart. J. of Math. (Oxford) (2) (1967), 369-84.

4. M. Kneser, Jahresbericht der Deutschen Mathm. Vereinigung Aufgabe 360, 58, 2 Abt. Heft 2, 1955.

5. C. Berge, Nombres de coloration de l'hypergraphe h-parti complet, this volume.

RECONSTRUCTION THEOREMS FOR INFINITE HYPERGRAPHS

Richard Rado, University of Reading

The substance of this note forms part of work carried out jointly with C. Berge. Capital letters denote sets, and $|A|$ denotes the cardinal of A. In this paper, the terms "hypergraph" and "family of sets" are used as synonyms, so that a hypergraph is a family

(1) $$H = (E_i : i \in M),$$

in which the sets M and E_i may be finite or infinite. If M is infinite then H is called an infinite hypergraph. It is convenient to use the abbreviations

$$E_I = \cup(i \in I)\, E_i \quad \text{for } I \subseteq M,$$

$$E_{[J]} = \cap(j \in J) E_j \quad \text{for } \emptyset \subset J \subseteq M,$$

and similarly for letters other than E. The hypergraph (1) is said to be strongly isomorphic to the hypergraph

(2) $$H' = (F_i : i \in M)$$

if there is a bijection $\varphi: E_M \to F_M$ such that

$$\varphi(E_i) = F_i \quad \text{for } i \in M.$$

Throughout this note we retain the notation (1) and (2). Strong isomorphism is denoted by the relation

(3) $$H \cong H'.$$

Weak isomorphism, denoted by the relation

(4) $$H \simeq H',$$

means that there is a bijection $\varphi: E_M \to F_M$ and a permutation

$$\pi: M \to M \text{ such that } \varphi(E_i) = F_{\pi(i)} \text{ for } i \in M.$$

In the special case when $|E_i|, |F_i| \leq 2$ for $i \in M$, we are dealing with graphs, and (4) becomes isomorphism between graphs. In the general case, H is called <u>reconstructible</u> if the condition

$$(E_i: i \in M - \{k\}) \cong (F_i: i \in M - \{k\}) \text{ for all } k \in M$$

implies (3). Reconstructibility, in the case of graphs, is not directly relevant for the wellknown Ulam conjecture as in the latter we are concerned with weak isomorphism whereas here reconstructibility requires strong isomorphism.

With H is associated the family of <u>boolean atoms</u>

$$(E_{J,M}: \emptyset \subset J \subseteq M),$$

where $E_{J,M} = E_{[J]} - E_{M-J}$ for $\emptyset \subset J \subseteq M$. The atoms of H' are denoted by $F_{J,M}$. The following result is well known.

<u>Lemma</u>. <u>(3) holds if and only if</u>

$$|E_{J,M}| = |F_{J,M}| \text{ for } \emptyset \subset J \subseteq M.$$

We shall now prove that (Theorem 1) corresponding to every non-empty index set M there exists a non-reconstructible hypergraph, and that (Theorem 2) every infinite hypergraph H is reconstructible if all its "edges" E_i are finite.

<u>Theorem 1</u>. <u>Let</u> $M \neq \emptyset$. <u>Then there are sets</u> E_i, F_i, <u>for</u> $i \in M$, <u>such that</u>

(<u>i</u>) $(E_i: i \in I) \cong (F_i: i \in I)$ <u>for</u> $I \subset M$,

(<u>ii</u>) $(E_i: i \in M) \not\approx (F_i: i \in M)$.

<u>Remark</u>. In view of the use of the symbol $\not\approx$, the proposition (ii) is, in fact, stronger than is required for H to be non-reconstructible.

We shall give two proofs of Theorem 1. The first proof is direct and gives explicitly the sets E_i and F_i. The second proof, though non-constructive in the case of an infinite M, proceeds more symmetrically and yields two hypergraphs which are important in other contexts.

First Proof. Choose $\mu \in M$. For every $i \in M$, choose a set D_i such that

$$0 < |D_\mu| \leq |D_i| \geq \aleph_0 \quad \text{for every} \quad i \in M - \{\mu\},$$

$$D_i \cap D_j = \emptyset \quad \text{for} \quad i, j \in M; \ i \neq j.$$

$$\text{Put} \quad E_\mu = D_M; \ F_\mu = D_{M - \{\mu\}},$$

$$E_i = F_i = D_i \quad \text{for} \quad i \in M - \{\mu\} .$$

Proof of (i). Let $k \in M$. It suffices to show that

$$(5) \qquad (E_i: \ i \in M - \{k\}) \cong (F_i: \ i \in M - \{k\}).$$

Case 1. $k \neq \mu$. Since $|D_\mu| \leq |D_k| \geq \aleph_0$ we have $|D_\mu \cup D_k| = |D_k|$ and there is a bijection

$$\psi: \ D_\mu \cup D_k \to D_k.$$

Define a function φ on D_M by putting

$$\varphi(x) = \begin{cases} \psi(x) & \text{for} \quad x \in D_\mu \cup D_k \\ x & \text{for} \quad x \in D_{M - \{\mu, k\}} . \end{cases}$$

Clearly, φ is injective, and

$$\varphi(E_\mu) = \varphi(D_\mu \cup D_k) \cup \varphi(D_{M - \{\mu,k\}})$$

$$= D_k \cup D_{M - \{\mu,k\}} = F_\mu.$$

Also, $\varphi(E_i) = F_i$ for $i \in M - \{\mu,k\}$.

Case 2. $k = \mu$. We put $\varphi(x) = x$ for $x \in D_{M - \{k\}}$. Then $\varphi(E_i) = F_i$ for

$i \in M - \{k\}$, and (5) follows.

Proof of (ii). Assume $H \cong H'$, so that there is a bijection $\varphi: E_M \to F_M$ and a permutation $\pi: M \to M$ such that $\varphi(E_i) = F_{\pi(i)}$ for $i \in M$. Consider the sets $S = \cup(i \in M)(E_i - E_{M - \{i\}})$; $T = \cup(i \in M)(F_i - F_{M - \{i\}})$. We have

$$|S| = |\cup(i \in M)(\varphi(E_i) - \varphi(E_{M - \{i\}}))|$$

$$= |\cup(i \in M)(F_{\pi(i)} - F_{M - \{\pi(i)\}})| = |T|.$$

On the other hand, $E_\mu - E_{M - \{\mu\}} = D_\mu$ and $E_i - E_{M - \{i\}} = F_j - F_{M - \{j\}} = \emptyset$ for $i \in M - \{\mu\}$; $j \in M$. Hence $|S| = |D_\mu| > 0 = |T|$, which contradicts (6). This proves Theorem 1.

Second proof. Put $\Omega = \{I: I \subseteq M\}$; $P = \{1, 3, 5, \ldots\}$; $d(I,J) = |I-J| + |J-I|$ for $I, J \in \Omega$. There is, by Zorn's lemma if M is infinite, a maximal set $\Omega' \subseteq \Omega$ such that $d(I, J) \notin P$ whenever $I, J \in \Omega'$. Put $\Omega'' = \Omega - \Omega'$, and define an equivalence relation on Ω by putting $I \equiv J$ whenever either $I, J \in \Omega'$ or $I, J \in \Omega''$. We now show that $I \equiv J$ implies $d(I, J) \notin P$. Otherwise there are sets $I'', J'' \in \Omega''$ such that $d(I'', J'') \in P$. Then, by the maximality of Ω', there are sets $I', J' \in \Omega'$ such that $d(I', I'') = d(J', J'') \in P$. Then $d(I', J')$ is finite and

$$d(I', J') \equiv d(I', I'') + d(I'', J'') + d(J'', J') \equiv 1 \pmod 2.$$

Thus we obtain the contradiction $d(I', J') \in P$.

Consider the hypergraphs

$$K(M) = (K_i(M): i \in M); \; L(M) = (L_i(M): i \in M)$$

where, for $i \in M$,

$$K_i(M) = \{I: i \in I \equiv M\}; \; L_i(M) = \{I: i \in I \not\equiv M\}.$$

We show that $K(M)$ and $L(M)$ are non-reconstructible. For simplicity we write

K, L, K_i, L_i instead of $K(M), L(M), K_i(M), L_i(M)$ respectively. It suffices to prove

$$(7) \qquad (K_i : i \in M - \{k\}) \cong (L_i : i \in M - \{k\}) \quad \text{for} \quad k \in M,$$

$$(8) \qquad |K_{[M]}| > 0 = |L_{[M]}|.$$

Clearly, (8) implies $K \not\cong L$ which, together with (7), provides the required proof. In order to describe the structure of K and L a little further, we shall also show that, for each $i \in M$,

$$(9) \qquad |K_i| = |L_i| = \begin{cases} 2^{|M| - 2} & \text{if } 2 \leq |M| < \aleph_0, \\ 2^{|M|} & \text{if } |M| \geq \aleph_0. \end{cases}$$

Proof of (7). Let $k \in M$. Put

$$f(X) = (X - \{k\}) \cup (\{k\} - X) \quad \text{for} \quad X \subseteq M.$$

Then $f(f(X)) = X$, and f permutes the subsets of M. Thus f is a bijection $\Omega \to \Omega$. Also, the definition of the relation $\equiv$ shows that $X \not\equiv f(X)$ for $X \in \Omega$.

Now consider any element i of $M - \{k\}$ and any set $I \in K_i$. Then $i \in I \equiv M$ and hence $i \in f(I) \not\equiv I \equiv M$; $f(I) \in L_i$; $f(K_i) \subseteq L_i$. Similarly, if $J \in L_i$ then $i \in J \not\equiv M$; $i \in f(J) \not\equiv J \not\equiv M$. Hence $f(J) \in K_i$; $f(L_i) \subseteq K_i$. Hence $L_i = f(f(L_i)) \subseteq f(K_i) \subseteq L_i$, and $f(K_i) = L_i$, which proves (7).

Proof of (8). We have $M \in K_i$ for all $i \in M$, so that $K_{[M]} \neq \emptyset$. On the other hand, if $I \not\equiv M$ and $I \subset M$, there is $j \in M - I$. Then $I \notin L_j$. Hence $I \notin L_{[M]}$. Since I was arbitrary, we have $L_{[M]} = \emptyset$, and (8) follows.

Proof of (9). Let $i \in M$. Then we can choose $k \in M - \{i\}$. By (7), $|K_i| = |L_i|$. Also, $K_i \cap L_i = \emptyset$; $|K_i \cup L_i| = |\{I : i \in I \subseteq M\}| = 2^{|M - \{i\}|}$. This implies (9).

Theorem 2. Let E_i be finite for all $i \in M$. Suppose that

(10) $$(E_i:\ i \in I) \cong (F_i:\ i \in I) \text{ for every finite } I \subseteq M.$$

Then $H \cong H'$.

Remark. The theorem shows that the hypergraph H is already reconstructible from its finite subhypergraphs, which is a stronger proposition than is needed for reconstructibility in the sense in which this term was introduced.

Proof. Since $|E_i| < \aleph_0$ we deduce from (10) that $|F_i| < \aleph_0$ for $i \in M$. For $I, J \subseteq M$ and $I \neq \emptyset$ we put $E(I, J) = E_{[I]} - E_J$; $F(I, J) = F_{[I]} - F_J$. We note that $E(I, M - I)$ is the atom $E_{I, M}$ of H, and similarly for H'. It follows from our definition that

(11) $$E(I, J) \supseteq E(I', J') \text{ if } I \subseteq I' \subseteq M \text{ and } J \subseteq J' \subseteq M.$$

Let $\emptyset \subset I_0 \subseteq M$. Choose $k \in I_0$. If

(12) $$k \in I \subseteq I_0;\ J \subseteq M - I_0;\ |I|,\ |J| < \aleph_0,$$

then $$|E(I, J)| \leq |E_k| < \aleph_0.$$

Hence there are finite sets $I_1, J_1 \subseteq M$ such that $k \in I_1 \subseteq I_0$; $J_1 \subseteq M - I_0$ and

(13) $$|E(I, J)| \geq |E(I_1, J_1)|$$

for all choices of sets I, J satisfying (12). By (11),

(14) $$E(I_1, J_1) \supseteq E(I_0, M - I_0).$$

We shall now show that there is equality in (14). Assume that there is an element

(15) $$x_0 \in E(I_1, J_1) - E(I_0, M - I_0).$$

If $x_0 \notin E_{[I_0]}$ then there is $r \in I_0$ such that $x_0 \notin E_r$. Then, by (11) and (13),

$$x_0 \notin E(I_1 \cup \{r\}, J_1) = E(I_1, J_1),$$

which contradicts (15). Hence $x_0 \in E_{[I_0]}$. Then, by (15), we have $x_0 \in E_{M - I_0}$, and therefore $x_0 \in E_s$ for some $s \in M - I_0$. Then, by (11) and (13),

$$x_0 \notin E(I_1, J_1 \cup \{ s \}) = E(I_1, J_1),$$

which is a contradiction. This proves that there is equality in (14). We now use (10) and the fact that the sets I_1 and J_1 are finite. This gives, together with (11),

$$|E(I_0, M - I_0)| = |E(I_1, J_1)| = |F(I_1, J_1)|$$
$$\geq |F(I_0, M - I_0)|.$$

By interchanging the roles of H and H' we find

$$|F(I_0, M - I_0)| \geq |E(I_0, M - I_0)|.$$

Thus corresponding atoms of H and H' have the same cardinal, and (3) follows from the Lemma.

NOTE ON A HYPERGRAPH EXTREMAL PROBLEM

M. Simonovits, Eotvos L. University, Budapest

Introduction. We shall consider 3-uniform hypergraphs "without loops or multiple edges." This means that we shall consider a set X which will be called the vertex-set of the hypergraph and a set of unordered triples from X called the triples of the hypergraph. The expression "without loops" means that each triple has 3 different elements and the expression "without multiple edges" means that each triple can occur at most once in the set of edges.

Problem. Let L be a family of hypergraphs. What is the maximum number of triples a hypergraph on n vertices can have if it does not contain a subhypergraph isomorphic to some members of L ? For a given finite or infinite family L, the problem asked above will be called an extremal problem; the maximum will be denoted by $\mathrm{ext}(n ; L)$; the members of L will be called sample hypergraphs, and the hypergraphs attaining the maximum generally are called extremal hypergraphs.

Definition. The r-pyramid $L_{r,t}$ based on a polygon of t vertices is the hypergraph defined on the $r + t$ vertices $x_1, \dots, x_r ; y_1, \dots, y_{t-1}, y_t, y_{t+1} = y_1$ and having the triples

$$(x_i, y_j, y_{j+1}) \quad (i=1, \dots, r ; j=1, \dots, t)$$

Further, L_r is the family of all the r-pyramids $L_{r,t}$, $t = 2, 3, \dots$.

Theorem. $\mathrm{ext}(n ; L_3) > (\frac{1}{6} + o(1))\, n^{8/3}$.

(This means that there are hypergraphs with n vertices and almost $\frac{1}{6} n^{8/3}$ triples which do not contain any 3-pyramid.)

Remarks. 1) In [1] we needed and proved the following lemma:

$$\mathrm{ext}(n ; L_r) = o(n^{3 - \frac{1}{r}}), \quad r = 1, 2, 3, \dots \tag{1}$$

This lemma was needed to prove that a 4-colour-critical hypergraph cannot have too many independent vertices. As a matter of fact, we needed this result only for $r = 2$. At the same time W. Brown, P. Erdos and V. T. Sos, [1] among some other hypergraph extremal theorems proved that if T is the family of sample hypergraphs obtained from the triangulations of the 3-sphere, then

$$\mathrm{ext}(n\ ;\ T) = O(n^{3-\frac{1}{2}}) \quad . \tag{2}$$

They used the fact that T contains L_2, proved (1) for $r = 2$, from which (2) followed trivially. Using a finite geometrical construction, they also proved that (2) is sharp, i.e. the exponent is the best possible. We used a so-called probabilistic argument to prove the weaker assertion that (1) is sharp for $r = 2$. The main purpose of this paper is to prove that (1) is sharp for $r = 3$ as well.

2) There is a result of Kovary, Turan and T. Sos [3] asserting that if $K_2(p\ ,\ q)$ is the complete bipartite graph with p and q vertices in its first and second classes respectively, then

$$\mathrm{ext}(n\ ;\ K_2(p\ ,\ q)\ ) = O(n^{2-\frac{1}{p}})\ ,\quad (p \leq q) \tag{3}$$

It can be conjectured that this result is the best possible; however, this is not proved except for $p = 1$, 2, and 3. If we can prove that (1) is sharp for $r = 4$, then it will follow that (3) is also sharp for $p = 4$. This suggests that it will be difficult to prove the sharpness of (1) for $r = 4$.

<u>The construction</u>. The construction given below will be based on the construction of W. G. Brown [4], showing that (3) is sharp for $p = 3$. In view of the second remark, this is not "surprising" at all. First we define the graph of Brown. Let p be an odd prime and the vertices of our graph be the points in the 3-dimensional affine space over the field $GF(p)$, i.e. over the field of residues mod p.

Let us join two points $\underline{x}$ and $\underline{y}$ by an edge if

$$\sum_{i=1}^{3} (x_i - y_i)^2 = a \tag{4}$$

where a is a quadratic residue or non-residue depending on whether p has the form $4s + 3$ or $4s + 1$, $a \neq 0$ and is constant for a given graph. According to a well-known theorem of Lebesgue [5, p. 325] the valences of each vertex will be $p^2 - p$ and, as Brown proves, this graph never contains a $K_2(3,3)$. Let us join 3 points $\underline{x}$, $\underline{y}$, $\underline{z}$ in the graph of Brown by a triple if

(5) $$\sum_{i=1}^{3} (x_i + y_i + z_i)^2 = a \ .$$

This hypergraph contains some 3-pyramids of very special "position". We omit a few triples and prove that the obtained hypergraph does not contain 3-pyramids. Let us denote the graph of Brown by B , the hypergraph defined above by A and let U be the hypergraph obtained from A by omitting all those vertices $\underline{x} = (x_1, x_2, x_3)$ for which

(6) at least one of x_1 , x_2 , x_3 vanishes .

Of course we omit all the triples at least one vertex of which has already been omitted. Since in U each coordinate can be chosen in $p - 1$ different ways, U has

(7) $$(p - 1)^3$$

vertices. For every edge $(\underline{x} , \underline{c})$ of B , there are $\frac{p-3}{2}$ pairs $(\underline{y} , \underline{z})$ such that $\underline{y} + \underline{z} = \underline{c}$, $\underline{y} \neq \underline{z}$, $\underline{y} \neq \underline{x}$, $\underline{z} \neq \underline{x}$. These $(\underline{x} , \underline{y} , \underline{z})$ triples will belong to A and each triple of A can be counted only 3 times this way, so A has $\frac{p^8}{6} + O(p^7)$ triples and each vertex of A has the valence $\frac{p^5}{2} + O(p^4)$. Since we omitted only $O(p^2)$ vertices, the number of triples in U is

(8) $$\frac{p^8}{6} + O(p^7) \ .$$

We shall prove that U does not contain 3-pyramids.

Let us suppose that the vertices $\underline{x}_i$ $(i = 1 , 2 , 3)$ and $\underline{y}_j$ $(j = 1,\dots,k)$ define an $L_{3,k}$ in U . The triples of this $L_{3,k}$ are the triples $(\underline{x}_i, \underline{y}_j , \underline{y}_{j+1})$. Let $\underline{w}_j = -(\underline{y}_j + \underline{y}_{j+1})$. According to (4) and (5) the vertices $\underline{x}_i$ are joined to the vertices $\underline{w}_j$ in the graph B . We know that the vertices $\underline{x}_i$ are all different and that the graph B does not contain a complete bipartite graph with 3 vertices in each class. Hence there are at most 2 different vertices among $\underline{w}_1 ,\dots,\underline{w}_k$. On the other hand

(9) $$-(\underline{w}_j - \underline{w}_{j-1}) = \underline{y}_j + \underline{y}_{j+1} - \underline{y}_{j-1} - \underline{y}_j = \underline{y}_{j+1} - \underline{y}_{j-1} \neq 0 \ .$$

This gives that the set $\{\underline{w}_j : j = 1, \dots, k\}$ has exactly 2 elements and every second element of it is the same. Therefore

(10) $$k = 2m \text{ and } \underline{w}_{2j} = \underline{u}, \ w_{2j-1} = \underline{v} \quad (j = 1, \dots, m), \ \underline{u} \neq \underline{v}.$$

Let us notice that from (9), (10) and $\underline{y}_{2m+1} = \underline{y}_1$ follows

$$0 = (\underline{y}_1 - \underline{y}_3) + (\underline{y}_3 - \underline{y}_5) + \dots + (\underline{y}_{2m-3} - \underline{y}_{2m-1}) + (\underline{y}_{2m-1} - \underline{y}_{2m+1}) = m(\underline{u} - \underline{v})$$

and, since $\underline{u} \neq \underline{v}$, m must be a multiple of p. Until now we considered the larger hypergraph A. It is easy to see that A can contain 3-pyramids. However, U cannot contain these subgraphs. Indeed, the vertices $\underline{w}_{2i+1}$ form an arithmetic progression of vectors with increment $\underline{u} - \underline{v}$. Hence at least one coordinate, e.g. the first of $\underline{u} - \underline{v}$, is different from 0; therefore, the first coordinates of the vectors $\underline{w}_{2j+1}$ form an arithmetic progression of residues mod p. Since the number of elements in this progression is at least p, at least one of them must be 0; hence the corresponding $\underline{w}_{2j+1}$ does not belong to U. Thus, (7) and (8) complete the proof of our theorem if $n = (p-1)^3$. Since $\text{ext}(n ; L)$ is monotone increasing in n and the primes are fairly dense among the integers (i.e. for every $\varepsilon > 0$ the interval $(n - \varepsilon n, n)$ contains a prime if n is large enough), our theorem follows for every n.

REFERENCES

1. Simonovits, M., On critical chromatic graphs, Studia Math. Hung. Acad. Sci. (to appear).

2. Brown, W. G., Erdös, P., and Sós, V. T., On the existence of triangulated spheres in 3-graphs and related problems, Studia Math. Hung. Acad. Sci. (to appear).

3. Kovary, T., Sós, V. T. and Turan, P., On a problem of K. Zarankiewicz, Coll. Math. 3 (1955) 50 - 57.

4. Brown, W. G., On graphs that do not contain a Thomsen graph, Canadian Math. Bull. 9 (1966) 281 - 285.

5. Dixon, L. E., Theory of numbers, Vol. II (Chelsea Publishing Company, 1952).

SUR UNE CONJECTURE DE V. CHVATAL

F. Sterboul, Université Paris-VI

I.- Introduction

Soit H un hypergraphe dont l'ensemble des arêtes est $\mathcal{E}$. H est dit héréditaire si, pour toute arête E de $\mathcal{E}$, tout sous-ensemble non-vide de E appartient à $\mathcal{E}$. Le rang de H est l'entier $\max_{E \in \mathcal{E}} |E|$. Le degré d'un sommet de H est le nombre d'arêtes contenant ce sommet. Le degré de H est le plus grand des degrés des sommets. Une clique de H est un sous-ensemble de $\mathcal{E}$ constitué d'arêtes se coupant deux à deux. Une clique peut être maximale (pour la relation d'inclusion des ensembles) ou maximum (aucune autre clique ne possède plus d'arêtes). Une étoile est une clique dont toutes les arêtes contiennent un sommet donné.

Pour tout hypergraphe H, le cardinal d'une clique maximum est au moins égal au degré de H (qui est le cardinal d'une certaine étoile de H). Chvatal a proposé la conjecture suivante :

. Conjecture : Pour un hypergraphe héréditaire H, le cardinal d'une clique maximum est égal au degré de H.

Cette conjecture est évidemment équivalente aux deux suivantes :

. Conjecture : Pour un hypergraphe héréditaire H, le cardinal de toute clique est au plus égal au degré de H.

. Conjecture : Pour un hypergraphe héréditaire H, il existe toujours une étoile appartenant à l'ensemble des cliques maximum .

Enoncée sous cette dernière forme, la conjecture conduit au problème plus général de la détermination des cliques maximum d'un hypergraphe héréditaire qui est résolu ici pour certains cas particuliers.

Dans (1), Chvátal a démontré sa conjecture pour une classe particulière d'hypergraphes héréditaires : les hypergraphes H tels que

- les sommets de H sont les entiers du segment $[1,n]$

- si E est une arête de H, F un ensemble de sommets de H, et s'il existe une injection de F dans E vérifiant $f(x) \geq x \quad \forall x \in F$, alors F est une arête de H.

Erdös, Chao Ko, Rado (3) ont montré que l'ensemble des cliques maximales coïncide avec l'ensemble des cliques maximum pour l'hypergraphe héréditaire complet (tel que tout ensemble non vide de sommets est une arête).

On démontre ici la conjecture de Chvátal pour plusieurs classes particulières d'hypergraphes héréditaires :

Théorème 1.- Soit H un hypergraphe héréditaire tel que

- Les arêtes maximales (pour l'inclusion) sont toutes de cardinal $h \geqslant 3$,
- L'intersection de tout couple d'arêtes maximales est de cardinal $\leqslant 1$,
- Il existe au moins 2 arêtes maximales d'intersection non vide,

alors toutes les cliques maximum sont des étoiles.

Remarque 1 : Ce théorème s'applique en particulier aux hypergraphes héritidaires dont les arêtes maximales sont des droites d'un plan projectif fini (problème posé par Chvátal dans (2)).

Remarque 2 : On démontre aisément la conjecture pour les hypergraphes héréditaires de rang 2, en remarquant que leurs cliques sont soit des étoiles, soit des triangles.

Théorème 2.- Dans un hypergraphe héréditaire de rang 3, le cardinal de toute clique est au plus égal au degré de l'hypergraphe, et les cliques maximum ont l'une des structures suivantes :

- Etoile
- {(1/2) , (1/3) , (2/3) , (1/2/3)}
- $\{(1/2/x_1),\dots,(1/2/x_p)$, $(1/3/x_1),\dots,(1/3/x_p)$, $(2/3/x_1),\dots,(2/3/x_p)$, (1/2) , (1/3) , (2/3), et éventuellement (1/2/3)} .
- {(1/2) , (1/3) , (1/4) , (1/2/3) , (1/2/4) , (1/3/4) , (2/3/4)}

(Les entiers représentent ici des sommets).

Remarque 3 : On voit de plus qu'avec les hypothèses du Théorème 2 l'existence de cliques maximum non étoiles implique des conditions très restrictives sur la structure de l'hypergraphe.

Théorème 3.- Soit H un hypergraphe héréditaire, et $E_1,\dots,E_p$ ses arêtes maximales (pour l'inclusion). On suppose qu'il existe un ensemble de sommets F tel que $E_i \cap E_j = F \ \forall i,j \ (i \neq j)$. Alors, le cardinal de toute clique est au plus égal au degré de H.

Remarque 4 : On détermine de plus dans la démonstration du Théorème 3 la structure de toutes les cliques maximum.

Théorème 4.- Soit H un hypergraphe héréditaire dont le nombre d'arêtes maximales est 3. On suppose que l'intersection de ces arêtes maximales est non vide. Alors le cardinal de toute clique est au plus égal au degré de H.

Remarque 5 : D'après le Théorème 3, on a le même résultat pour les hypergraphes héréditaires ayant 1 ou 2 arêtes maximales.

Démonstration du Théorème 1 :

Soit C une clique maximale non étoile de H. Comme C est maximale, les arêtes maximales de C sont de cardinal h. Soit M le nombre d'arêtes maximales de C. Pour tout sommet x, soit $f(x)$ le nombre d'arêtes maximales de C contenant x. Pour toute arête maximale E_i de C, soit k_i le cardinal de l'ensemble $\{x : x \in E_i , f(x) \geq 2\}$.

Soit F le maximum de la fonction f. Soient $x_1,\dots,x_{k_i}$ les sommets de E_i où la valeur de f est au moins égale à 2. Alors :

$$M = 1 + (f(x_1)-1)+\dots+(f(x_{k_i})-1) \leq 1 + k_i(F-1)$$

d'où :

$$k_i \geq M-1/F-1 \qquad (I).$$

Un sous-ensemble de E_i appartient à C ssi il contient $\{x_1,\dots x_{k_i}\}$ et, comme C n'est pas une étoile, une arête de C n'appartient qu'à une seule arête maximale. Il en résulte que :

$$|C| = \sum_{i=1}^{M} 2^{h-k_i}$$

De plus, le degré de H est au moins égal à $1+F(2^{h-1} - 1)$. La démonstration du théorème se ramène à montrer que :

$$|C| \leq F(2^{h-1}-1).$$

Comme C n'est pas une étoile, on a $k_i \geq 2$, pour tout i, d'où

$|C| \leq M.2^{h-2}$. Il en résulte que le théorème est vrai dans le cas où :

$$M \leq F(2-2^{-(h-2)}).$$

. Soit alors $n \geq 1$ tel que

$$F(2^n)(1-2^{-(h-1)}) < M \leq F(2^{n+1})(1-2^{-(h-1)}) \qquad \text{(II)}$$

En combinant (I) avec la première inégalité de (II) on obtient :

$$k_i > 2^n + (F-1)^{-1}(2^n-1-2^n.F.2^{-(h-1)}) \qquad \text{(III)}$$

En supposant ici que le deuxième terme de la somme n'est pas négatif, on déduit de (III): $k_i \geq 2^n+1 \geq n+2$, d'où $|C| \leq M.2^{h-(n+2)}$, et le résultat cherché s'obtient alors en appliquant la deuxième inégalité de (II)

Il reste à déterminer le signe de $A = 2^n-1-2^n.F.2^{-(h-1)}$. On remarque que, du fait que C n'est pas une étoile, on a $F \leq h$.

<u>Cas</u> $h \geq 4$: alors $h.2^{-(h-1)} \leq 1/2$, d'où $A \geq 0$.

<u>Cas</u> $h=3$: si $F=2$, on a encore $A \geq 0$. Le seul cas restant est donc le cas où $h=3$ et $F=3$, et il faut alors montrer que l'inégalité $M \geq 5$ entraîne $|C| \leq 9$. (I) entraîne $M \leq 7$, et si $|C|$ était supérieur ou égal à 10, il y aurait au moins 3 arêtes de cardinal 2 dans C, et on vérifie aisément que cette hypothèse est incompatible avec l'existence dans C de 5 arêtes de cardinal 3.

<u>Démonstration du Théorème 2</u> :

Soit H un hypergraphe héréditaire de rang 3 et C une clique maximale non étoile. On distingue plusieurs cas qui seront eux-mêmes subdivisés.

<u>Cas 1</u> : Toutes les arêtes de C sont de cardinal 3.

<u>Cas 1.1</u> : L'intersection de tout couple d'arêtes de C est de cardinal 1. Le théorème est alors démontré en appliquant le théorème 1 à l'hypergraphe héréditaire dont les arêtes maximales sont les arêtes de C.

<u>Cas 1.2</u> : En désignant les sommets de H par des entiers, il existe deux arêtes de la forme : (1/2/3) et (1/2/4). Si toutes les arêtes de C contenaient le sommet 1 ou le sommet 2, $C \cup \{(1/2)\}$ serait une clique et C ne serait pas maximale. Donc il existe i arêtes de la forme : (3/4/x) , $i \neq 0$, $x \geq 5$. Les autres arêtes de C sont de la forme :

(1/3/4),(2/3/4),(1/2/x),(1/3/x),(1/4/x),(2/3/x),(2/4/x),(1/x/y),(2/x/y)
$y > x \geq 5$, en nombres respectifs :

b , c , d , e , f , g , h , j , k.

On a :

$$|C| = 2+b + c + d + e + f + g + h + i + j + k$$

$b \leq 1$, $c \leq 1$ et, du fait que $i \neq 0$, $d \leq 1$.

<u>Cas 1.2.1</u> : $j \geq 2$ (ou de façon symétrique $k \geq 2$). Puisqu'il existe une arête (3/4/5), les arêtes (1/x/y) sont de la forme (1/5/y) et, comme $j \geq 2$, on a : $g \leq 1$, $h \leq 1$ et $i=1$.

Alors :

$$|C| \leq 6 + b + d + e + f + j + k$$

et, $d(x)$ désignant le degré dans H du sommet x,

$$d(1) \geq 6 + b + d + e + f + j + (j+1)$$

où le terme $j+1$ provient de la considération des arêtes (1/5) et (1/y) incluses dans (1/5/y).

<u>Cas 1.2.1.1.</u> : $k \leq j$. Le théorème est alors démontré car $|C| < d(1)$.

<u>Cas 1.2.1.2</u> : $k \geq j+1$. Alors $k \geq 3$, et il en résulte :

$e \leq 1$ et $f \leq 1$.

Alors :

$$|C| \leq 6 + c + d + g + h + k + j$$

$$d(2) \geq 6 + c + d + g + h + k + (k+1) \quad \text{et} \quad |C| < d(2).$$

<u>Cas 1.2.2.</u> : $j \leq 1$ et $k \leq 1$. Alors :

$$|C| \leq 4 + b + d + e + f + j + g + h + i$$

$$d(1) \geq 6 + b + d + e + f + j + \sup(e,f)$$

où le terme $\sup(e,f)$ provient de la considération des arêtes (1/x) incluses dans (1/3/x) ou (1/4/x).

Le théorème est alors démontré si : $g + h + i \leq 1 + \sup(e,f)$

ou de façon symétrique si : $e + f + i \leq 1 + \sup(g,h)$.

D'autre part,

$$|C| \leq 4 + b + c + e + g + i + f + h + d$$

$$d(3) \geq 5 + b + c + e + g + i + \sup(e,g, i)$$

Le théorème est alors démontré si : $f + h + d \leq \sup(e,g,i)$
ou de façon symétrique si : $e + g + d \leq \sup(f,h,i)$

Il suffit donc de démontrer que le système d'inégalités suivant n'admet pas de solution, compte tenu des propriétés des variables :

(I) $g + h + i \geq 2 + \sup(e,f)$

(II) $e + f + i \geq 2 + \sup(g,h)$

(III) $f + h + d \geq 1 + \sup(e,g,i)$

(IV) $e + g + d \geq 1 + \sup(f,h,i)$

<u>Cas 1.2.2.1</u> : $i \geq 2$. Ceci entraîne : $d=0$. D'où d'après (III) : $f+h \geq 3$. Ceci entraîne $f \geq 2$ ou $h \geq 2$. Prenons, par exemple $f \geq 2$: on en déduit $g = 0$, puis avec (IV) : $e \geq 3$, qui lui-même entraîne $h=0$. Finalement, (III) et (IV) se réduisent à $f \geq 1+e$ et $e \geq 1+f$: impossible.

<u>Cas 1.2.2.2.</u> : $i=1$. De (I) et (II) on déduit successivement : $g+h \geq 1$, $\sup(g,h) \geq 1$, $e+f \geq 2$ et $g+h \geq 2$.

<u>Cas 1.2.2.2.1</u> : Un des 4 nombres e,f,g,h est au moins égal à 2. Par exemple $e \geq 2$. Alors, $h=0$, d'où $g \geq 2$, d'où $f=0$. (I) et (II) se réduisent alors à : $g \geq 1+e$ et $e \geq 1+g$: impossible.

<u>Cas 1.2.2.2.2.</u> : $e = f = g = h = 1$. Alors :

$$d(1) \geq 9 + b + d + j$$
$$|C| \leq 7 + b + d + j + c + k.$$

Le théorème est alors démontré si : $c + k \leq 1$
ou de façon symétrique si : $b + j \leq 1$.

Le dernier cas serait donc : $b = c = j = k = 1$, mais on remarque qu'il ne peut avoir lieu, car l'arête (2/3/4) ne coupe pas une arête (1/x/y).

<u>Cas 2</u> : C contient une arête de cardinal 2 : (1/2). Alors, $C = \{(1/2)\} \cup C_1 \cup C_2 \cup D$ où C_1 (resp. C_2) est l'ensemble des arêtes de C de la forme $(1/x)$ ou $(1/x/y)$ (resp. $(2/x)$ ou $(2/x/y)$) avec $y > x > 2$. D est l'ensemble des arêtes $(1/2/x)$. Pour un sommet x, notons $d'(x)$ le degré de x dans l'hypergraphe héréditaire engendré par $C-D$. On remarque alors que dans les cas où on peut prouver que $d'(1) > |C-D|$ ou $d'(2) > |C-D|$, il en résulte $d(1) > |C|$ ou $d(2) > |C|$ et C n'est pas une clique maximum.

Cas 2.1. : C-D ne contient que des arêtes de cardinal 2. Comme C est une clique non étoile, C-D est une clique non étoile, donc un triangle : $\{(1/2), (1/3), (2/3)\}$. Posons p=0 ou 1 suivant que (1/2/3) n'appartient pas ou appartient à D, et $q = |D - \{(1/2/3)\}|$. Alors : $|C| = 3 + p + q$ et $d(1) \geq 3 + p + 2q$. Donc, C ne peut être une clique maximum à l'exception du cas où

$$C = \{(1/2), (1/3), (2/3), (1/2/3)\}$$

Cas 2.2. : C_1 contient une arête de cardinal 3 : (1/3/4). On rencontrera souvent dans la suite le cas où l'intersection de toutes les arêtes de $C_1 \cup C_2$ est non vide. Examinons le ici :

Cas 2.2.1. : Alors, posons a(resp. b) = 1 ou 0 suivant que (1/3) (resp. (2.3)) appartient ou non à C_1 (resp. C_2). On suppose que le sommet 3, par exemple, appartient à toutes les arêtes de $C_1 \cup C_2$. Soit $(1/3/x_1),\ldots,(1/3/x_p)$ l'ensemble des arêtes de cardinal 3 de C_1, et, pour C_2, $(2/3/y_1),\ldots,(2/3/y_q)$. (q = 0 si cet ensemble est vide).

Alors, $|C-D| = 1 + a + b + p + q$

$d'(1) \geq 3 + 2p$

$d'(2) \geq 2 + b + 2q$

Le seul cas n'interdisant pas à C d'être une clique maximum est donc : a = b = 1 et p = q. Dans ce cas,

$$C_1 = \{(1/3), (1/3/x_1),\ldots,(1/3/x_p)\}$$

$$C_2 = \{(2/3), (2/3/y_1),\ldots,(2/3/y_p)\}$$

Posons r=1 ou 0 suivant que (1/2/3) appartient ou non à D et $D - \{(1/2/3)\} = \{(1/2/z_1),\ldots,(1/2/z_s)\}$ s = 0 si cet ensemble est vide.

$$|C| = 3 + 2p + r + s \ , \ d(1) \geq 3 + 2p + r + s$$

et remarquons que l'on peut encore ajouter une unité au minorant de d(1) si il existe un i tel que $z_i \notin \{x_1,\ldots x_p\}$. Pour que C soit de cardinal maximum, il faut donc que :

$$\{z_1,\ldots z_s\} \subset \{x_1,\ldots,x_p\} \qquad \{z_1,\ldots,z_s\} \subset \{y_1,\ldots,y_p\}.$$

Mais alors, $d(3) \geq 3 + 3p + r$, et finalement, pour que C soit une clique maximum, il faut s=p et la structure de C est :

$$C = \{(1/2),(1/3),(2/3),(1/3/x_1),\ldots,(1/3/x_p),(2/3/x_1),\ldots,(2/3/x_p)$$
$$(1/2/x_1),\ldots,(1/2/x_p) \text{ et éventuellement } (1/2/3)\}$$

Revenons à l'étude du cas 2.2.

<u>Cas 2.2.2.</u> : (1/3) et (1/4) appartiennent à C_1.

Alors, $C_2 = \{(2/3/4)\}$.

<u>Cas 2.2.2.1.</u> : C_1 contient une arête de cardinal 3 autre que (1/3/4).

Alors, $d'(1) \geq 3 + |C_1|$ $|C-D| = 2 + |C_1|$

et C ne peut être de cardinal maximum.

<u>Cas 2.2.2.2.</u> : $C_1 = \{(1/3),(1/4),(1/3/4)\}$ $C_2 = \{(2/3/4)\}$. Posons g(resp. h) = 1 ou 0 suivant que (1/2/3) (resp. (1/2/4)) appartient ou non à D ; et $p = |D-\{(1/2/3),(1/2/4)\}|$.

Alors, $|C| = 5 + g + h + p$ $d(1) \geq 5 + g + h + 2p$

$d(3) \geq 6 + g$ $d(4) \geq 6 + h$

Le seul cas où C peut être une clique maximum est donc : p=0 g = h = 1 et $C = \{(1/3),(1/4),(1/2),(1/3/4),(2/3/4),(1/2/3),(1/2/4)\}$.

<u>Cas 2.2.3.</u> : $(1/3) \in C_1$ et $(1/4) \notin C_1$.

Alors, $d'(1) \geq 3 + |C_1|$ $|C-D| = 1 + |C_1| + |C_2|$

Si $|C_2| \leq 1$, C ne peut être de cardinal maximum. Supposons donc que $|C_2| \geq 2$.

<u>Cas 2.2.3.1.</u> : C_2 contient une arête de cardinal 2. C'est alors nécessairement (2/3), et on rentre alors dans le cas 2.2.1.

<u>Cas 2.2.3.2.</u> : C_2 ne contient que des arêtes de cardinal 3.

Alors, $d'(2) \geq 5 + |C_2|$. Si $|C_1| \leq 3$, C ne peut être une clique maximum, on suppose donc $|C_1| \geq 4$.

Si $|C_2| \geq 3$, on voit aisément que le sommet 3 appartient à toutes les arêtes de C_1 et C_2 et on rentre dans le cas 2.2.1.

Si $|C_2| = 2$, alors $|C-D| = 3 + |C_1|$, $d'(1) > 3 + |C_1|$ et C ne peut être de cardinal maximum.

<u>Cas 2.2.4</u> : $(1/3) \notin C_1$ et $(1/4) \notin C_1$

Alors, $d'(1) \geq 4 + |C_1|$, d'où C ne peut être une clique maximum si $|C_2| \leq 2$. On suppose donc $|C_2| \geq 3$.

<u>Cas 2.2.4.1.</u> : C_2 ne contient que des arêtes de cardinal 3.

Alors, $d'(2) \geq 5 + |C_2|$, et pour que C soit de cardinal maximum il faut : $|C_1| \geq 4$. Dans ce cas, l'existence dans C_1 d'une arête de de cardinal 2 est incompatible avec l'hypothèse $|C_2| \geq 3$. Donc C_1 a au moins 4 arêtes de cardinal 3, et C_2 au moins 3. La seule configuration possible est alors celle du cas 2.2.1.

Cas 2.2.4.2. : C_2 contient une seule arête de cardinal 2.

Alors, $d'(2) \geq 4 + |C_2|$, et on peut supposer $|C_1| \geq 3$. Dans ce cas, que C_1 contienne ou non une arête de cardinal 2, on rentre dans le cas 2.2.1.

Cas 2.2.4.3. : C_2 contient 2 arêtes de cardinal 2 : (2/3) et (2/4).

Alors, $C_1 = \{(1/3/4)\}$. Si (2/3/4) appartient à C_2, on obtient un cas analogue au cas 2.2.2. en échangeant les numéros des sommets 1 et 2. On suppose donc que $(2/3/4) \notin C_2$.

Soit p (resp. q) le nombre d'arêtes de C_2 de la forme (2/3/x) (resp. (2/4/x)). $|C-D| = 4 + p + q$ et $d'(2) \geq 4 + p + q + \sup(p,q)$, et comme $|C_2| \geq 3$, $\sup(p,q)$ est non nul, donc C ne peut être une clique maximum.

Pour les démonstrations suivantes on utilisera les lemmes :

Lemme 1.- Soit X un ensemble fini et G un sous-ensemble de $\mathcal{P}(X)$ tel que : $E \in G \Longrightarrow \overline{E} \in G$. Si C est une clique de l'hypergraphe G alors $|C| \leq |G|/2$.

Démonstration : Si E est une arête de C, $\overline{E}$ n'est pas une arête de C. Donc G-C contient au moins $|C|$ arêtes.

Lemme 2.- Soit X un ensemble fini et G un sous-ensemble de $\mathcal{P}(X)$ tel que : $E \in G \Longrightarrow \overline{E} \in G$. Soient D_1 et D_2 deux sous-ensembles de G tels que : $E_1 \in D_1$, $E_2 \in D_2 \Longrightarrow E_1 \cap E_2 \neq \emptyset$. Alors : $|D_1| + |D_2| \leq |G|$.

Démonstration : Si E appartient à D_1, $\overline{E}$ n'appartient pas à D_2 donc $|D_2| \leq |G| - |D_1|$.

Démonstration du Théorème 3 :

Soient $E_i (1 \leq i \leq p)$ les arêtes maximales de l'hypergraphe H. On pose $|E_i| = h_i$, $F = E_i \cap E_j$ $(i \neq j)$, $|F| = h$.

Le degré de H est $2^{h_1-1} + \ldots + 2^{h_p-1} - (p-1)\, 2^{h-1}$.

Soit C une clique maximale de H :

Premier cas : Supposons qu'il existe une arête E de C et un indice i tels que $E \subset E_i - F$. Alors, toutes les arêtes de C seront contenues dans E_i et la remarque 5 s'applique ici.

<u>Second cas</u> : On suppose désormais que toutes les arêtes de C coupent F. On partitionne C de la façon suivante :

1). Arêtes de C contenues dans F : en appliquant le lemme 1, le nombre de ces arêtes est au plus 2^{h-1}.

2). Arêtes de C contenant strictement F et différentes de E_i $1 \leq i \leq p$: le nombre de ces arêtes est au plus :

$$(2^{h_1-h} - 2) + \ldots + (2^{h_p-h} - 2).$$

3). Arêtes de C appartenant à un des ensembles C_i :

$$C_i = \{E : E \in C \, , \, E_i - F \subset E \subset E_i\}$$

En appliquant alors le lemme 2 avec $X=F$, $G = \mathcal{P}(F)$, $D_1 = \{E \cap F \, , \, E \in C_i\}$ $D_2 = \{E \cap F \, , \, E \in C_j\}$, on a $|C_i| + |C_j| \leq 2^h$ $(i \neq j)$. D'où $|C_1| + \ldots + |C_p| \leq p \, 2^{h-1}$.

4). Arêtes de C appartenant à un des ensembles F_i :

$$G_i = \{E : E \subset E_i \, , \, E \not\subset F \, , \, E \not\subset E_i - F \, , \, E \not\supset F \, , \, E \not\supset E_i - F\}$$

$$F_i = G_i \cap C.$$

En appliquant alors le lemme 1 avec $X = E_i$ et $G = G_i$, on a : $|F_i| \leq |G_i|/2$ donc :

$$|F_i| \leq 2^{h_i-1} - 2^h - 2^{h_i-h} + 2.$$

Au total on a donc $|C| \leq 2^{h_1-1} + \ldots + 2^{h_p-1} - (p-1)2^{h-1}$.

Pour que C soit une clique maximum il faut et il suffit que les bornes soient atteintes pour chacun des ensembles de la partition précédente. Une telle clique C est donc la réunion

1). D'une clique maximum de l'hypergraphe $\mathcal{P}(F)$: C_o.

2). De l'ensemble de toutes les arêtes F telles que pour un indice i : $F \subset E \subset E_i$.

3). Des ensembles $C_i = \{E : E_i - F \subset E \subset E_i \, , \, E \cap F \in C_o\}$.

En effet si $E \in C$ et $E_i - F \subset E \subset E_i$, $E \cap F$ doit couper toutes les arêtes de la clique maximum C_o, donc $E \cap F \in C_o$.

4). Des ensembles G_i :

$$G_i = \{E : E \subset E_i \,,\ E \not\supset F \,,\ E \not\supset E_i - F \,,\ E \not\subset F \,,\ E \not\subset E_i - F \,,\ E \cap F \in C_o\}$$ ×

pour la même raison qu'en 3).

Démonstration du Théorème 4 :

Soient E_1, E_2, E_3 les arêtes maximales de l'hypergraphe H.

On pose $F_k = E_i \cap E_j$ $(i \neq j \neq k)$, $F = E_1 \cap E_2 \cap E_3$, $|E_i| = h_i$, $|F_i| = k_i$, $|F| = h$.

Le degré de H est $2^{h_1-1} + 2^{h_2-1} + 2^{h_3-1} - 2^{k_1-1} - 2^{k_2-1} - 2^{k_3-1} + 2^{h-1}$

Soit C une clique de H ;

Premier cas : On suppose qu'il existe au moins une arête de C contenue dans F. On partitionne alors C de la façon suivante :

1). Arêtes de C contenues dans un des ensembles F_i et non contenues dans F : en appliquant le Théorème 3 à ce système d'arêtes, on obtient que le nombre des arêtes considérées est au plus

$$2^{k_1-1} + 2^{k_2-1} + 2^{k_3-1} - 2^h - 2^{h-1}$$

2). Arêtes de C appartenant à un des ensembles G_1, G_2, G_3 :

$$G_1 = \{E : E \subset E_1,\ E \not\subset F,\ E \not\supset F,\ E \not\subset E_1 - F,\ E \not\supset E_1 - F,\ E \not\subset F_j,\ E \not\supset E_1 - F_j\ (j=2,3)\}$$

G_2 et G_3 sont définis de façon analogue. D'après le lemme 1 avec $X=E_1$ et $G=G_1$, le nombre d'arêtes de $C \cap G_1$ est au plus

$$2^{h_1-1} + 2^h - 2^{k_2} - 2^{k_3} - 2^{h_1-h} + 2^{k_2-h+1} + 2^{k_3-h+1} - 2.$$

3). Arêtes de C appartenant à un des ensembles H_1, H_2, H_3 :

$$H_1 = \{E : E \subset E_1 \,,\ E \supset F,\ E \not\supset E_1 - F_j,\ E \not\subset F_j\ \ (j = 2,3)\}.$$

H_2 et H_3 sont définis de façon analogue. Alors,

$$|H_1| \leq 2^{h_1-h} - 2^{k_2-h+1} - 2^{k_3-h+1} + 2$$

Il reste à considérer les arêtes de C contenues dans F et les arêtes E de C telles que pour deux indices i et j $(i \neq j)$:

$$E_i - F_j \subset E \subset E_i .$$

On répartit ces arêtes dans les groupes de deux sous-ensembles suivants :

4) $J_1 = \{E : E \subset E_1,\ E_1 - F \subset E\}$

$J_2 = \{E : E \subset E_2,\ E_2 - F_3 \subset E,\ E \cap F_3 \subset F\}$

5) $J_3 = \{E : E \subset E_2,\ E_2 - F \subset E\}$

$J_4 = \{E : E \subset E_1,\ E_1 - F_3 \subset E,\ E \cap F_3 \subset F\}$

6) $J_5 = \{E : E \subset E_3,\ E_3 - F \subset E\}$

$J_6 = \{E : E \subset E_2,\ E_2 - F_1 \subset E,\ E \cap F_1 \subset F\}$

7) $J_7 = \{E : E \subset E_1,\ E_1 - F_2 \subset E,\ E \cap F_2 \subset F\}$

$J_8 = \{E : E \subset E_3,\ E_3 - F_2 \subset E,\ E \cap F_2 \subset F\}$

8) $J_9 = \{E : E \subset E_3,\ E_3 - F_1 \subset E,\ E \cap F_1 \subset F\}$

$J_{10} = \{E : E \subset F\}$

9) $L_1 = \{E : E \subset E_2,\ E_2 - F_3 \subset E,\ E \cap F_3 \not\subset F,\ E \cap F_3 \not\supset F_3 - F\}$

$L_2 = \{E : E \subset E_1,\ E_1 - F_3 \subset E,\ E \cap F_3 \not\subset F,\ E \cap F_3 \not\supset F_3 - F\}$

10) $L_3 = \{E : E \subset E_2,\ E_2 - F_1 \subset E,\ E \cap F_1 \not\subset F,\ E \cap F_1 \not\supset F_1 - F\}$

$L_4 = \{E : E \subset E_3,\ E_3 - F_1 \subset E,\ E \cap F_1 \not\subset F,\ E \cap F_1 \not\supset F_1 - F\}$

11) $L_5 = \{E : E \subset E_1,\ E_1 - F_2 \subset E,\ E \cap F_2 \not\subset F,\ E \cap F_2 \not\supset F_2 - F\}$

$L_6 = \{E : E \subset E_3,\ E_3 - F_2 \subset E,\ E \cap F_2 \not\subset F,\ E \cap F_2 \not\supset F_2 - F\}$

Considérons le groupe 4). En appliquant le Lemme 2 avec $X=F$, $G = \mathcal{P}(F)$, $D_1 = \{E \cap F,\ E \in J_1 \cap C\}$ et $D_2 = \{E \cap F,\ E \in J_2 \cap C\}$ on a

$$|J_1 \cap C| + |J_2 \cap C| \leq 2^h.$$

On traite de même les groupes 5) à 8).

Considérons le groupe 9). En appliquant le Lemme 2 avec $X=F_3$ $G = \{E' : E' \subset F_3, E' \not\subset F, F_3-F \not\subset E'\}$, $D_1 = \{E \cap F_3, E \in L_1 \cap C\}$ et $D_2 = \{E \cap F_3, E \in L_2 \cap C\}$, on a :

$$|L_1 \cap C| + |L_2 \cap C| \leq 2^{k_3} - 2^{h+1}$$

On traite de même les groupes 10) et 11). Au total on a donc :

$$|C| \leq 2^{h_1-1} + 2^{h_2-1} + 2^{h_3-1} + 2^{h-1} - 2^{k_1-1} - 2^{k_2-1} - 2^{k_3-1}.$$

Second cas : On suppose qu'il n'existe pas d'arête de C contenue dans F. En plus des arêtes considérées dans le premier cas, il existe alors des arêtes de C ne coupant pas F. On répartit ces arêtes dans les sous-ensembles suivants :

1). $M_1 = \{E : E \subset E_1-F, E \not\subset F_3, E \not\subset F_2, E_1-F_2 \not\subset E, E_1-F_3 \not\subset E\}$.

M_2 et M_3 définis de façon analogue.

On remarque alors que $E \in M_1 \Longrightarrow E_1-E \in H_1$ et $E \in M_1 \cap C \Longrightarrow E-E_1 \notin H_1 \cap C$. Il en résulte que $|M_1 \cap C| + |H_1 \cap C| \lneq |H_1|$ (L'adjonction dans le second cas d'arêtes de M_1 estcompensée par la disparation d'arêtes de H_1).

2). Les arêtes restantes sont celles qui contiennent un ensemble du type E_i-F_j, elles appartiennent alors à l'un des ensembles J_1 à J_9 ou ou L_1 à L_6 et sont traitées comme dans le premier cas.

REFERENCES

1. V. Chvátal, Intersecting subfamilies of hereditary families, this volume.

2. V. Chvátal, D.A. Klarner, D.E. Knuth, Selected combinatorial research problems. (Problème 25). Stanford Univ. C.S. 72-292.

3. P. Erdös, Chao Ko, R.Rado, Intersection theorems for systems of finite sets. Quart. J. Math. Oxford (2), 12 (1961).

ON THE CHROMATIC NUMBER OF THE DIRECT PRODUCT OF HYPERGRAPHS

F. Sterboul, Université Paris VI

1. INTRODUCTION

The direct product $H_1 \times H_2$ of two hypergraphs H_k, $k = 1,2$, with set of vertices X_k and set of edges E_k, is the hypergraph whose set of vertices is $X_1 \times X_2$ (cartesian product) and set of edges $\{A_i \times B_j\}$ where A_i (resp. B_j) ranges over E_1 (resp. E_2).

For any hypergraph H, with set of vertices X, a subset of X is independent if it contains no edge of H. $\beta(H)$, the weak stability number of H, is the cardinality of the maximum independent subset of H. A coloring of H is a mapping from X into a finite set (of "colors") such that the vertices with the same color form an independent subset. $\chi(H)$, the weak chromatic number of H, is the minimum number of colors in a coloring of H (Erdös-Hajnal [6]).

K_m^r denotes the complete uniform hypergraph of rank r with m vertices, $H(m,n,r,s)$ the hypergraph $K_m^r \times K_n^s$, whose stability and chromatic numbers are $\beta(m,n,r,s)$ and $\chi(m,n,r,s)$. For the sake of brevity, set : $H(m,n) = H(m,n,2,2)$, $\beta(m,n) = \beta(m,n,2,2)$, $\chi(m,n) = \chi(m,n,2,2)$.

Zarankiewicz [12] asked for the value of $\beta(m,n,r,s)$. This problem has given rise to many results (for an extensive bibliography see Guy [8]). $\chi(m,n,r,s)$ was studied as a finite "polarized partition relation" (Erdös-Rado [7], Chvátal [3]). The purpose of this paper is to give some results about $\chi(m,n)$ and $\chi(m,n,r,s)$ (§ 2 to 6), and to extend the methods and formulae used for Zarankiewicz's problem and polarized partition relations to the general case of the product of two hypergraphs.

2. LOWER BOUNDS FOR χ

For any hypergraph H, one has (C. Berge [1])

(2.1) $$\chi(H).\beta(H) \geq n$$

where n is the number of vertices of H.

Reiman (11) has shown that :

(2.2) $$\beta(p,q) \leq (p + \sqrt{p^2+4pq(q-1)})/2$$

(2.1) and (2.2) lead obviously to :

(2.3) $$\chi(p,q) \geq 2pq.(p + \sqrt{p^2+4pq(q-1)})^{-1}$$

Equality holds in (2.2) when $p = m^2 + m$ and $q = m^2$, or when $p = q = m^2 + m + 1$. (m is a power of a prime number). In this case, (2.3) gives :

(2.4) $$\chi(m^2+m \ , \ m^2) \geq m$$

(2.5) $$\chi(m^2+m+1 \ , \ m^2+m+1) \geq m+1$$

(formula (2.5) is in Chvátal [3]).

(2.6) $$\chi(m^2 \ ,m^2) \geqslant m$$

We prove in §3 and §4 that there is actually equality in (2.4), (2.5) and (2.6).

3. EXACT VALUES FOR χ. THE AFFINE PLANE

Let P be the affine place over a field F of cardinal m. P has m^2 points and $m^2 + m$ lines. Let Q be the set of points, L the set of lines.

A coloring of $H(m^2+m \ , \ m^2)$ with m colors is equivalent to a mapping $Q \times L \xrightarrow{f} F$ such that the equalities

$$f(q_1,l_1) = f(q_1,l_2) = f(q_2,l_1) = f(q_2,l_2)$$

imply that either $q_1 = q_2$ or $l_1 = l_2$.

A line l has equation : $Y + aX + b = 0$ or $X + c = 0$. A point q has coordinates (x,y). Let us define f by

$f(q,l) = y + ax + b$ in the first case

$f(q,l) = x+c$ in the second case.

It is easily seen that f yields a suitable coloring of $H(m^2+m\ ,\ m^2)$ so that, taking account of inequalities (2.4) and (2.6), we have

$$(3.1) \qquad \chi(m^2+m\ ,\ m^2) = m$$

$$(3.2) \qquad \chi(m^2,m^2) = m$$

Remark :

$f(q,l) = 0$ iff q is a point of l, so that the matrix whose entries are $f(q,l)$ generalizes the incidence matrix of the affine plane.

4. EXACT VALUES FOR χ. THE PROJECTIVE PLANE

The projective plane P' over F is defined by adding to the affine plane P a new line l_o (line at infinity) and $m+1$ new points (at infinity) on this line. Each line now contains $m+1$ points. Two lines, distinct from l_o, are parallel if they contain the same point at infinity. Let $q_i (1 \leq i \leq m+1)$ be the points of l_o, and C_i the corresponding classes of parallels. The set of points of P' is $Q' = Q \cup Q_o$ (where Q_o is the set of points of l_o) and the set of lines is $L' = L \cup \{l_o\}$.

Let us add a new element γ to the set F. As in §3, a coloring of $H(m^2 + m + 1,\ m^2 + m + 1)$ is equivalent to a mapping $Q' \times L' \overset{g}{\to} F \cup \{\gamma\}$, satisfying the same condition as in §3. We shall use a latin square A, of order $m+1$, whose entries $a(i,j)$ are elements of $F \cup \{\gamma\}$ and such that $a(i,i) = 0$ (this last condition allows the remark of §3 to remain true).

Let us define the mapping g as follows :

- the restriction of g to $Q \times L$ is equal to f
- $g\ (Q \times \{l_o\}) = \{\gamma\}$
- $g(Q_o \times \{l_o\}) = \{0\}$
- $g(q_i,l) = a(i,j)$ if $l \in C_j \subset L$ and $q_i \in Q_o$

It is easily seen that this yields a coloring of $H(m^2+m+1\ ,\ m^2+m+1)$ with $m+1$ colors. Hence, using (2.5), we have

$$(4.1) \qquad \chi(m^2+m+1\ ,\ m^2+m+1) = m+1$$

Figure 1 is an example of a coloring for $m=4$.

Figure 1 : Coloring of $H(m^2+m,m^2)$ and $H(m^2+m+1, m^2+m+1)$ for $m=4$.

	Points at infinity					Q															
1_o	0	0	0	0	0	4	4	4	4	4	4	4	4	4	4	4	4	4	4	4	4
	0	1	2	3	4	0	0	0	0	1	1	1	1	2	2	2	2	3	3	3	3
	0	1	2	3	4	1	1	1	1	0	0	0	0	3	3	3	3	2	2	2	2
	0	1	2	3	4	2	2	2	2	3	3	3	3	0	0	0	0	1	1	1	1
	0	1	2	3	4	3	3	3	3	2	2	2	2	1	1	1	1	0	0	0	0
Class of parallels	1	0	4	2	3	3	2	1	0	3	2	1	0	3	2	1	0	3	2	1	0
	1	0	4	2	3	2	3	0	1	2	3	0	1	2	3	0	1	2	3	0	1
	1	0	4	2	3	1	0	3	2	1	0	3	2	1	0	3	2	1	0	3	2
	1	0	4	2	3	0	1	2	3	0	1	2	3	0	1	2	3	0	1	2	3
	2	3	0	4	1	3	2	1	0	2	3	0	1	1	0	3	2	0	1	2	3
	2	3	0	4	1	2	3	0	1	3	2	1	0	0	1	2	3	1	0	3	2
	2	3	0	4	1	1	0	3	2	0	1	2	3	3	2	1	0	2	3	0	1
	2	3	0	4	1	0	1	2	3	1	0	3	2	2	3	0	1	3	2	1	0
	3	4	1	0	2	3	2	1	0	1	0	3	2	0	1	2	3	2	3	0	1
	3	4	1	0	2	2	3	0	1	0	1	2	3	1	0	3	2	3	2	1	0
	3	4	1	0	2	1	0	3	2	3	2	1	0	2	3	0	1	0	1	2	3
	3	4	1	0	2	0	1	2	3	2	3	0	1	3	2	1	0	1	0	3	2
	4	2	3	1	0	3	2	1	0	0	1	2	3	2	3	0	1	1	0	3	2
	4	2	3	1	0	2	3	0	1	1	0	3	2	3	2	1	0	0	1	2	3
	4	2	3	1	0	1	0	3	2	2	3	0	1	0	1	2	3	3	2	1	0
	4	2	3	1	0	0	1	2	3	3	2	1	0	1	0	3	2	2	3	0	1

<u>Remark</u> :

When m is not the power of a prime, there is no field of cardinal m. Nevertheless, if $p_1^r \ldots p_k^s$ is the decomposition of m into a product of prime powers, and if $d = p_1^r$ is the smallest of the factors $p_1^r \ldots p_k^s$, there is a coloring of $H(m^2+m, d.m)$ with m colors. Indeed, let R be a ring of cardinal m, and S a subset of R with cardinality d, such that for any distinct x and

x' in S, $x-x'$ has an inverse in R. (This is always possible). Then one can use the method of §3 where the set of points with coordinates (x,y) is limited to those for which $x \in S$.

5. ASYMPTOTIC BEHAVIOUR OF χ

Formula (2.3) yields

(5.1) $$\liminf \frac{\chi(n,n)}{\sqrt{n}} \geq 1$$

Let p and q be two consecutive prime numbers such that $p^2 \leq n \leq q^2$, then (3.2) implies $p \leq \chi(n,n) \leq q$ hence $\chi(n,n) \leq \frac{q}{p}.n$. Since the quotient of consecutive primes tends to 1, we have

(5.2) $$\limsup \frac{\chi(n,n)}{\sqrt{n}} \leq 1,$$

and this together with (5.1) implies

(5.3) $$\lim \frac{\chi(n,n)}{\sqrt{n}} = 1 .$$

6. $\chi(p,q,r,s)$

The paragraphs above emphasized the case in which $r = s = 2$. Chvátal gave several formulas and a review of previous results for the general case in [3].

One can deduce from the results of §4 a coloring of $H((m^2+m+1)(r-1), (m^2+m+1)(s-1), r,s)$ with $m+1$ colors by repeating the pattern of the coloring of $H(m^2+m+1 , m^2+m+1)$ $(r-1) \times (s-1)$ times. Thus, for m the power of a prime,

(6.1) $$\chi((m^2+m+1)(r-1) , (m^2+m+1)(s-1),r,s) \leq m+1$$

In the same way

(6.2) $$\chi(m^2(r-1) , m^2(s-1),r,s) \leq m$$

(6.3) $$\chi(m^2(r-1),(m^2+m)(s-1),r,s) \leq m$$

Actually, there is equality in these formulas when $r = 2$. The proof is easy by combining (2.1) with the Hylten-Cavallius formula [9]

(6.4) $$\beta(p,q,2,s) \leq q/2 + ((s-1)qp(p-1) + (q^2/4))^{1/2}$$

Thus

(6.5) $\quad \chi(m^2, m^2(s-1), 2, s) = \chi(m^2, (m^2+m)(s-1), 2, s) = m$

(6.6) $\quad \chi(m^2+m+1,\ (m^2+m+1)(s-1),\ 2, s) = m+1$

Remark :

$$\beta(m^2, (m^2+m)(s-1),\ 2, s) = (s-1)(m^3+m^2)$$

$$\beta(m^2+m+1,\ (m^2+m+1)(s-1),\ 2, s) = (s-1)(m+1)(m^2+m+1)$$

As in §5, one may deduce asymptotic results from (6.5) :

(6.7) $$\lim_{n \to \infty} \frac{\chi(n, n(s-1), 2, s)}{\sqrt{n}} = 1$$

Remark :

Guy [8] has noted that $\beta(n,n,r,n-r+1) = n^2-n$. Hence in $H(n,n,r,n-r+1)$ the complement of a maximum independent set has cardinality n, so is also independent if $r < n$. Hence

$$\chi(n,n,r,n-r+1) = 2$$

7. THE STABILITY NUMBER OF A DIRECT PRODUCT

The stability number of a direct product was studied by C. Berge and Simonovits (this volume) from an other point of view, We want here to generalize the method used in the Zarankiewicz problem by Kövari, Sós, Turán [10], Reiman [11], Hylten-Cavallius [9], Znam [13], Znam and Guy [14]. Most of their formulas are special cases of the formulas given in this section.

In all that follows H(resp. H') will be a hypergraph with set of vertices X (resp. Y), numbers of stability, of vertices, of edges β, n, m (resp. β', n', m'). Moreover, u will be the minimum cardinal of an edge of H.

To H is associated a function f, that one may call the stability function, as follows : $f(k)$ is the minimum number of edges of a section-hypergraph of H, with k vertices, that is, $f(k)$ is the minimum number of edges that may be contained in a subset of X of cardinality k. Note that $\beta = \sup \{k : f(k) = 0\}$. This function is related to Turan's numbers for hypergraphs.

Theorem 7.1 $$f(k) \geq \frac{k}{k-u} f(k-1)$$

Proof :

Let H_k be a section-hypergraph of H with k vertices and f(k) edges. Any section-hypergraph of H_k with k-1 vertices has at least f(k-1) edges ; any edge of H_k is included in at most k-u section-hypergraphs of H_k with k-1 vertices ; there are k subhypergraphs of H_k with k-1 vertices. Hence, the sum of the numbers of edges of all the section-hypergraphs of H_k with k vertices is greater than k.f(k-1) and less than (k-u).f(k).

Corollary 7.2. We have

$$f(k) \geq \frac{k(k-1)\dots(k-u+1)}{(\beta+1)(\beta)\dots(\beta-u+2)} f(\beta+1), \quad k \geq \beta+1$$

Corollary 7.3 We have

$$f(k) \geq f(k-1)+1 \quad \text{and} \quad f(k) \geq f(\beta+1) + k-1-\beta$$

Now, let S be a subset of X × Y, and for any y in Y let l_y be the cardinality of $S \cap (X \times \{y\})$, and s the cardinality of S : $s = \sum_{y \in Y} l_y$. If

$$(7.4) \qquad \sum_{y \in Y} f(l_y) > \beta' m$$

then, by the pigeon-hole principle, there exists an edge E of H, and β'+1 vertices $y_j (0 \leq j \leq \beta')$ of H' such that $E \times \{y_j\}$ is included in S. Since $\{y_o, \dots, y_{\beta'}\}$ cannot be an independent set of H', it follows that, if condition (7.4) is fulfilled, there exists an edge F of H' such that E × F is contained in S, thus S cannot be an independent set of H × H'.

Let us now look for sufficient conditions on S in order that (7.4) be fulfilled.

Let p be the cardinality of $\{y : l_y \geq \beta + 1\}$

and

$$t = \sum_{l_y \geq \beta+1} l_y$$

then

$$(7.5) \qquad s \leq t + (n'-p)\beta)$$

Now by (7.3) and (7.5) :

$$\sum_{y \in Y} f(l_y) = \sum_{l_y \geq \beta+1} f(l_y) \geq p(f(\beta+1)-1) + t-p\beta$$

$$\geq s - n'\beta \quad \text{by (7.3)}.$$

Hence if $s > n'\beta + m\beta'$ condition (7.4) is fulfilled, and we have

Theorem 7.6 We have

$$\beta(H \times H') \leq n'\beta + m\beta'.$$

There is equality in (7.6) in the particular case of Čulik's formula (4) :

$$\beta(p,q,r,s) = (r-1)q + (s-1)\binom{p}{r} \text{ if } q \geq (s-1)\binom{p}{r}.$$

Let us now use (7.2) instead of (7.3) :

$$\sum_{y \in Y} f(l_y) \geq \sum_{l_y \geq \beta+1} \frac{l_y(l_y-1)\dots(l_y-u+1)}{(\beta+1)(\beta)\dots(\beta-u+2)} f(\beta+1)$$

The convexity of the function $x(x-1)\dots(x-u+1)$ and (7.5) then imply :

(7.7)
$$\sum_{y \in Y} f(l_y) \geq p \frac{(t/p)((t/p)-1)\dots((t/p)-u+1)}{(\beta+1)(\beta)\dots(\beta-u+2)} f(\beta+1)$$

$$\geq p \frac{(((s-n'\beta)/p)+\beta)(((s-n'\beta)/p)+\beta-1)\dots(((s-n'\beta)/p)+\beta-u+1)}{(\beta+1)(\beta)\dots(\beta-u+2)} f(\beta+1)$$

Let us first assume that $s > n'\beta$ in order that $p \geq 1$. Elementary calculus shows then that the minimum in (7.7) is reached for $p = n'$, if $s > n'(\gamma+\beta)$ where γ is the positive root of the equation :

$$\frac{1}{x} = \frac{1}{x+\beta} + \dots + \frac{1}{x+\beta-u+1}$$

(If $\beta = u-1$, that is, $H = K_n^u$, take $\gamma = 0$)

(A useful upper bound for γ is $\frac{\beta}{u-1} - \frac{1}{2}$)

This gives the

Lemma 7.8 If $\frac{(s/n')((s/n')-1)\dots((s/n')-u+1)}{(\beta+1)(\beta)\dots(\beta-u+2)} > \frac{\beta' m}{n' t(\beta+1)}$ and $s > n'(\gamma+\beta)$, then S contains an edge of $H \times H'$.

We first deduce from (7.8) the

Theorem 7.9. If $n' \geq \frac{\beta' m}{f(\beta+1)}$, then $\beta(H \times H') \leq n'(\beta+\sup(\gamma,1))$.

Using now the inequality

$(s/n')((s/n')-1)\ldots((s/n')-u+1) \geq ((s/n')-u+1)^u$ we have

Theorem 7.10. We have

$$\beta(H \times H') \leq \sup\left[n'(\gamma+\beta)\ ;\ (u-1)n' + n'^{+1-1/u}\left(\frac{\beta' m(\beta+1)(\beta)\ldots(\beta-u+2)}{f(\beta+1)}\right)^{1/u}\right]$$

This contains Hylten-Cavallius formula [9] (see also [10]) :

$$\beta(n,n',r,s) \leq (r-1)n'+(s-1)^{1/r}\, n'^{\,1-1/r}\, n.$$

Using lemma 7.8 in the case when $u=2$ we have

Theorem 7.11. If $u=2$, then

$$\beta(H \times H') \leq \sup\left[n'(\gamma+\beta)\ ;\ \frac{1}{2}.\left(n' + \sqrt{n'^2 + \frac{4\,\beta'\beta(\beta+1)mn'}{f(\beta+1)}}\right)\right]$$

This contains the Reiman formula [1] and the Hylten-Cavallius formula [9]:

$$\beta(n,n',2,s) \leq \frac{1}{2}\left(n' + \sqrt{n'^2 + 4(s-1)\, n(n-1)n'}\right)$$

where there is equality for certain values of n,n' (cf. §6).

8. ON THE CHROMATIC NUMBER OF A PRODUCT

By using inequality (2.1) one can deduce from any upper bound of β (for instance from those in §7) a lower bound for χ.

Using Chvátal's remark [3] that if the vertices of $H \times H'$ are colored with k colors there exists at least $n.n'/k$ vertices with the same color, one can deduce from lemma 7.8. the

Theorem 8.1. If $n/k \geq \gamma+\beta$ and if

$$n'.\frac{n}{k}\left(\frac{n}{k}-1\right)\ldots\left(\frac{n}{k}-u+1\right) f(\beta+1) > \beta' m(\beta+1)(\beta)\ \ldots\ (\beta-u+2)$$

then $\chi(H \times H') > k$

We refer the reader to [2] for other interesting formulas.

REFERENCES

1. C. Berge, Graphes et hypergraphes, Dunod, 1970.
2. C. Berge, M. Simonovits : The coloring numbers of the product of two hypergraphs, this volume.
3. Chvátal, On finite polarized partition relations . Can. Math. Bull. 12, n° 3, 1969, 321-326.
4. Čulik, Teilweise Losung eines verallgemeinerten Problems von K. Zarankiewicz. Annales Polonici Math., n° 3, 1956, 165-168.
5. P. Erdös, Some remarks on the theory of graphs. Bull. Amer. Math. Soc., 53, 1947, 292-299.
6. P. Erdös, A. Hajnal, On chromatic number of graphs and set-systems. Acta Math. Acad. Sc. Hungaricae, 17, 1966, 61-99.
7. P. Erdös, R. Rado, A partition calculus in set theory. Bull. Amer. Math. Soc., 62, 1956, 427-489.
8. R.K. Guy, The Many Facets of Graph Theory, Lecture Notes in Math, n° 110, Springer-Verlag, 129-148.
9. C. Hylten-Cavallius, On a Combinatorial problem. Colloq. Math., 6, 1958, 59-65.
10. T. Kövari, V.T. Sós, P. Turán, On a problem of K. Zarankiewicz. Colloq Math. 3, 1954, 50-57.
11. I. Reiman, Uber ein problem von K. Zarankiewicz. Acta Math. Acad. Sc. Hungaricae, 9, 1958, 269-279.
12. K Zarankiewicz, P. 101, Colloq. Math., 2, 1951, 301.
13. S. Znam, On a Combinatorial problem of K. Zarankiewicz. Colloq. Math. 11, 1963, 81-84.
14. R.K. Guy, S. Znam, A problem of Zarankiewicz. Recent Progress in Combinatorics. (W.T. Tutte ed.) 1969, 237-243.

EVERY DIRECTED GRAPH HAS A SEMI-KERNEL

V. Chvátal, Stanford University

L. Lovász, Vanderhilt University

In a directed graph, the distance $d(u,v)$ from a vertex u to a vertex v is the number of edges in the shortest directed path from u to v. It is well-known that every tournament has a vertex u such that $d(u,v) \leq 2$ for all v ; in fact, any vertex of largest outdegree is such a vertex [2] . This generalizes as follows :

Theorem : *In a directed graph G, there is always a set S of vertices such that*

(i) $d(u,v) \geq 2$ *whenever* $u,v \in S$ *and* $u \neq v$,

(ii) *given any* $v \notin S$ *there is an* $u \in S$ *with* $d(u,v) \leq 2$.

Proof. By induction on the number of vertices of G. Let w be a vertex of G ; let G' be the subgraph of G induced by $\{u:d(w,u)\geq 2\}$. By the induction hypothesis, there is a set S' which works for G'. If $d(u,w) \leq 1$ for some $u \in S'$, we set $S = S'$; otherwise we set $S = S' \cup \{w\}$. Obviously, S has the required properties.

REMARK. A set S satisfying (i) and such that

(iii) given any $v \notin S$ there is an $u \in S$ with $d(v,u) \leq 1$

is called a *kernel* (cf. [1]). Not every directed graph has a kernel.

REFERENCES

1. C. Berge, *Graphs and Hypergraphs*, North Holland, Amsterdam 1973, Chapter 14. Kernels and Grundy functions.

2. H.G. Landau, On dominance relations and the structure of animal societies, III ; the condition for a score structure, *Bull. Math. Biophys.* 15 (1955), 143-148.

ELEMENTARY STRONG MAPS AND TRANSVERSAL GEOMETRIES*

T.A. Dowling, The Ohio State University
D.G. Kelly, University of North Carolina at Chapel Hill

1. PRELIMINARIES

We consider throughout pregeometries [5] (or matroids) defined on a fixed finite set X. The rank function of a pregeometry G on X is denoted r_G. The rank of G is r(G), equal to $r_G(X)$. For $A \subseteq X$, $\overline{A}^G$ is the G-closure of A, and the geometric lattice of G-closed subsets, or G-flats, ordered by inclusion, is denoted by L(G).

If H, G are two pregeometries on X and every G-flat is an H-flat (equivalently, if $\overline{A}^G \supseteq \overline{A}^H$ for every $A \subseteq X$), then G is a quotient of H, and the identity map on X extends to a strong map [5,7], the canonical surjection, denoted $H \rightarrow G$. All strong maps considered here, except where stated otherwise, are of this type. If G is a quotient of H, then $r(G) \leq r(H)$ with equality holding iff G = H. The difference r(H) - r(G) is the order of $H \rightarrow G$. In

* This research was supported in part by the Air Force Office of Scientific Research under Contract n° AFOSR-68-1415. A portion of this work was undertaken while the authors attended the National Science Foundation Advanced Science Seminar on Combinatorial Theory at Bowdoin College, Brunswick, Me. in the summer of 1971.
We omit detailed proofs here, as the full paper will appear elsewhere.

particular, if H is the free geometry $\mathcal{B}$ on X, in which every subset of X is closed (and independent), $H \to G$ is the closure map of G, of order $|X|-r(G)$, the nullity $n(G)$ of G. If $G = \mathcal{B}^*$, the (rank zero) dual of $\mathcal{B}$, in which X is the only closed set, $H \to G$ has order $r(H)$, the rank of H.

We call a strong map of order zero trivial. A strong map of order one is elementary. Every nontrivial strong map $H \to G$ of order n admits an elementary factorization

$$H = G_0 \to G_1 \to G_2 \to \dots \to G_{n-1} \to G_n = G,$$

where $G_{i-1} \to G_i$ is elementary, $1 \leq i \leq n$. Such a factorization is not unique, but a canonical one is the Higgs factorization [5,7] in which G_{i-1} is the Higgs lift of G_i in H, $1 \leq i \leq n$.

A single-element extension [4,5] of a pregeometry H on X is a pregeometry K on $X \cup e$, where $e \notin X$, such that $r(H) = r(K)$ and the restriction (subgeometry) of K to X is H. By the strong map factorization theorem of Higgs [7] (see also Brylawski [3]), an elementary strong map $H \to G$ may be factored as an injection $H \to K$ into a single-element extension, followed by a contraction $K \to G$ of the added element e. The map is trivial if e is a loop in the extension K, and elementary otherwise. The extension K on $X \cup e$ is unique. There is consequently a one-one correspondence between proper single-element extensions K (those in which e is not a loop) of H and elementary quotients G of H.

Single-element extensions are in turn specified by modular cuts [4,5] of H. A modular cut of H is a nonempty order filter M in L(H) such that if A, B are in M and each covers their meet $A \wedge B$, then $A \wedge B$ is in M. M is proper if it is a proper subset of $L(H)$. The generators of a modular cut M are its minimal members. Every principal filter, i.e. one with a single generator, is a modular cut. A proper modular cut M of H determines an elementary quotient G of H (via $H \to K \to G$) by $L(G) = L(H) - S$, where S is the support of M, consisting of the H-flats not in M but covered by a (unique) member of M. Latticially, the strong map $H \to G$ fixes all flats of H not in S, and takes each flat of S to the unique member of M covering it. The rank function of G is

$$(1.1) \qquad r_G(A) = \begin{cases} r_H(A) - 1, & \overline{A}^H \in M \\ r_H(A), & \text{otherwise.} \end{cases}$$

If $M = L(H)$ is improper, then $G = H$, and $H \to G$ is trivial.

To specify an elementary (or trivial) strong map $H \to G$ we shall use the following notation. Let $\mathcal{S}$ be any set of subsets of X such that the set of H-flats containing one or more members of $\mathcal{S}$ is a modular cut of H. Then if G is the corresponding elementary quotient, we write $H \overset{\mathcal{S}}{\to} G$. We could of course take $\mathcal{S}$ to be the set of generators of M, but we shall not require that the members of $\mathcal{S}$ be closed in H. Then $H \overset{\mathcal{S}}{\to} G$ is trivial iff $\overline{\emptyset}^H$, the set of loops of H, contains a member of $\mathcal{S}$.

If the modular cut M defining an elementary (or trivial) strong map $H \to G$ is principal, we call $H \to G$ a principal map. Such a map may be denoted $H \overset{E}{\to} G$ where E is any set spanning the generator of M. A pregeometry P is principal if its closure map $\mathcal{B} \to P$ admits an elementary factorization into principal maps:

$$(1.2) \qquad \mathcal{B} = P_0 \overset{E_1}{\to} P_1 \overset{E_2}{\to} P_2 \to \cdots \to P_{n-1} \overset{E_n}{\to} P_n = P$$

We write $P = P(E_1, E_2, \dots, E_n)$, or briefly $P(\underline{E})$, in this case, where $\underline{E} = (E_1, E_2, \dots, E_n)$. Such pregeometries were first investigated by Brown [2], who called them "F-products".

Let $\underline{E} = (E_1, E_2, \dots, E_m)$ be a list of m subsets of X. A transversal of $\underline{E}$ is an m-subset $\{x_1, x_2, \dots, x_m\}$ of X such that $x_i \in E_i$, $1 \leq i \leq m$. A partial transversal of $\underline{E}$ is a transversal of a sublist of $\underline{E}$. By the defect version [8] of P. Hall's "marriage theorem", a subset A of X contains a partial transversal of size n of $\underline{E}$ iff for every $I \subseteq \{1,2,\dots,m\}$,

$$(1.3) \qquad |A \cap \bigcup_{i \in I} E_i| \geq |I| - (m-n).$$

The partial transversals of $\underline{E}$ are the independent sets of a pregeometry T on X, the transversal pregeometry of Edmonds and Fulkerson [6]. We write $T = T(E_1, E_2, \dots, E_m)$, or simply $T(\underline{E})$. The list $\underline{E}$ is a presentation of the pregeometry T.

2. RANK AND CLOSURE FOR AN ELEMENTARY FACTORIZATION

Let $H \to G$ be a nontrivial strong map of order n and let

$$(2.1) \qquad H = G_0 \overset{\delta_1}{\nrightarrow} G_1 \overset{\delta_2}{\nrightarrow} G \to \dots \to G_{n-1} \overset{\delta_n}{\nrightarrow} G_n = G$$

be an elementary factorization of $H \to G$. By (1.1), for $A \subseteq X$,

$$r_H(A) - r_G(A) = |\{i | \exists E_i \in \delta_i, E_i \subseteq \overline{A}^{G_{i-1}}\}|.$$

Define a set function f by

$$(2.2) \qquad f(A) = |\{i | \exists E_i \in \delta_i, E_i \subseteq A\}|.$$

The following may then be readily shown.

Proposition 2.1 Let A be a subset of X. Then $0 \leq f(A) \leq r_H(A) - r_G(A)$, and $f(A) = r_H(A) - r_G(A)$ if A is a G-flat.

Proposition 2.2 For any subset A of X,

$$r_G(A) = r_H(\overline{A}^G) - f(\overline{A}^G) = \min_{B \supseteq A} \{r_H(B) - f(B)\}.$$

Corollary 2.2.1 $r_G(A) = r_H(A)$ iff for each $B \supseteq A$, $f(B) \leq r_H(B) - r_H(A)$.

Corollary 2.2.2 A is G-independent iff A is H-independent and for each $B \supseteq A$, $f(B) \leq r_H(B) - |A|$.

Corollary 2.2.3 If $H = \mathcal{B}$ in (2.1), then

$$r_G(A) = |A| - f(\overline{A}^G) = \min_{B \supseteq A} \{|B| - f(B)\},$$

and A is G-independent iff for each $B \supseteq A$, $f(B) \leq |B - A|$.

Proposition 2.3 A subset A of X is a G-flat iff for each $B \supset A$,

$$f(B) - f(A) < r_H(B) - r_H(A).$$

Corollary 2.3.1 If $H = \mathcal{B}$ in (2.1), A is a G-flat iff for each $B \supset A$,

$$f(B) - f(A) < |B-A|.$$

3. ELEMENTARY STRONG MAPS AND PARTIAL TRANSVERSALS

Let G be a pregeometry of nullity n on X, and consider a factorization

$$(3.1) \qquad \mathcal{B} = G_0 \overset{\delta_1}{\nrightarrow} G_1 \overset{\delta_2}{\nrightarrow} G_2 \rightarrow \cdots \overset{\delta_n}{\nrightarrow} G_n \overset{\delta_{n+1}}{\nrightarrow} \cdots G_{m-1} \overset{\delta_m}{\nrightarrow} G_m = G$$

of the closure map of G into n elementary and m-n trivial maps. If $\mathcal{S}$ defines the trivial map on a pregeometry, then $\mathcal{S}$ defines the trivial map on any of its quotients, so we may assume without loss of generality that $\mathcal{S}_1, \mathcal{S}_2, \ldots, \mathcal{S}_n$ define elementary maps and $\mathcal{S}_{n+1}, \ldots, \mathcal{S}_m$ define trivial maps. Thus $G_n = G_{n+1} = \ldots = G_m = G$, and we may apply the results of section 2 to the sequence

$$(3.2) \qquad B = G_0 \overset{\mathcal{S}_1}{\to} G_1 \overset{\mathcal{S}_2}{\to} G_2 \to \cdots \overset{\mathcal{S}_n}{\to} G_n = G,$$

with f defined by (2.2). Let $\underline{\mathcal{S}}$ denote the product set $\mathcal{S}_1 \times \mathcal{S}_2 \times \ldots \times \mathcal{S}_n \times \ldots \times \mathcal{S}_m$ and denote an arbitrary member of $\underline{\mathcal{S}}$ by $\underline{E} = (E_1, E_2, \ldots, E_n, \ldots, E_m)$.

Theorem 3.1 Let G be a pregeometry of nullity n on X and let (3.1) be a factorization of the closure map of G into elementary and trivial maps. Then a subset of X is independent in G iff its complement contains a partial transversal of size n of every $\underline{E}$ in $\underline{\mathcal{S}}$.

By Cor. 2.2.3, X - A is G-independent iff $f(B) \le |B - A|$ for every $B \subseteq A$, and A contains a partial transversal of every $\underline{E}$ in $\underline{\mathcal{S}}$ iff (1.3) holds for every $I \subseteq \{1,2,\ldots,n,\ldots,m\}$ and every $\underline{E}$ in $\underline{\mathcal{S}}$. The proof is completed by showing the equivalence of these conditions.

Theorem 3.2 The nullity n of G is the minimum over all $\underline{E}$ in $\underline{\mathcal{S}}$ of the size of a maximum partial transversal of $\underline{E}$.

By Theorem 3.1 it suffices to show that for some $\underline{E}$ in $\underline{\mathcal{S}}$, the size of a maximum partial transversal of $\underline{E}$ is exactly n. By Cor. 2.2.3, $|\bar{\emptyset}^G| = f(\bar{\emptyset}^G)$, and so there exists $I \subseteq \{1,2,\ldots,n\}$ with $|I| = |\bar{\emptyset}^G|$ and $E_i \in \mathcal{S}_i$, $i \in I$, such that E_i contains $\bar{\emptyset}^G$. Also for $j = n+1, \ldots, m$, there exists $E_j \in \mathcal{S}_j$ such that E_j contains

$\overline{\phi}^G$. But then

$$|\bigcup_{i\in I} E_i \cup \bigcup_{j=n+1}^{m} E_j| \geq |I| - (m-n),$$

so no $\underline{E}$ in $\underline{\mathcal{E}}$ containing E_i, $i \in I$, and $E_{n+1},\ldots,E_m$ can have a partial transversal of size exceeding n.

On taking each $\mathcal{E}_i$ to be a singleton, we obtain (Brown [2])

Theorem 3.3 *Let $\underline{E} = (E_1,E_2,\ldots,E_m)$ be a list of subsets of X. Then the principal pregeometry $P(\underline{E})$ and the transversal pregeometry $T(\underline{E})$ are dual pregeometries.*

Corollary 3.3.1 *If $\underline{F} = (F_1,F_2,\ldots,F_m)$ is any permutation of the members of $\underline{E} = (E_1,E_2,\ldots,E_m)$, then $P(\underline{F}) = P(\underline{E})$.*

The following corollary is well-known in transversal theory.

Corollary 3.3.2 *A transversal pregeometry T of rank n may be presented by a list of n sets.*

Given a presentation of T by a list of $m \geq n$ sets, say $T = T(E_1,E_2,\ldots,E_m)$, let $P = T^*$. Then the closure map of the principal pregeometry P can be factored as in (3.1), with $\mathcal{E}_i = \{E_i\}$, and by deleting the $m - n$ trivial maps we obtain a presentation of T in terms of the n E_i's defining elementary maps.

Bondy and Welsh [1] define a presentation $(E_1,E_2,\ldots,E_n)$ of a transversal pregeometry T of rank n to be *minimal* if no element can be deleted from any E_i to yield another presentation, and *maximal* if no element can be added to any E_i to yield another presentation. Given a presentation $(E_1,E_2,\ldots,E_n)$ of T, its dual $P = T^*$ is principal of nullity n, and so the closure map of P admits an elementary factorization (1.2). The set E_n may be replaced in this factorization by any set E_n' with the same P_{n-1}-closure as E_n. In particular, if we let E_n' be the P_{n-1}-closure of E_n, then E_n' is the generator of the principal modular cut defining the elementary strong map $P_{n-1} \to P$, and hence is also P-closed. By Cor. 3.3.1, any set E_i may be exchanged with E_n to yield another elementary factorization of $\mathcal{B} \to P$, hence all E_i may be replaced by their P-closures. This yields a maximal presentation $(E_1', E_2', \ldots, E_n')$ of T in which every E_i' is closed in P, i.e. in which every $X - E_i'$ is isthmus-free in T. We thus obtain the following result, (Bondy and Welsh [1], Thm. 3).

<u>Theorem 3.4</u> <u>A transversal pregeometry T of rank n may be maximally presented by a list of n flats of T*.</u>

If instead we let E_n' be a basis of the P_{n-1}-closure of E_n, then E_n' is a circuit in P, since an elementary map truncates the restrictions to the cut generators by (1.1). Again by Cor. 3.3.1 it follows that T may be minimally presented by a family of circuits of P, or equivalently (Bondy and Welsh [1], Thm 1).

<u>Theorem 3.5</u> <u>A transversal geometry T of rank n may be minimally presented by a list of n bonds of T.</u>

4. BASIS INTERSECTIONS

We now discard trivial maps and consider an elementary factorization (3.2) of the closure map of a pregeometry G of nullity n. Theorems 4.1 and 4.1* below follow immediately from the results of section 3.

Theorem 4.1 Let G be a pregeometry of nullity n on X. Then there exists a family $\mathcal{P}$ of principal pregeometries of nullity n on X such that a subset of X is G-independent iff it is P-independent for every P in $\mathcal{P}$.

The desired family is $\mathcal{P} = \{P(\underline{E}) | \underline{E} \in \underline{\mathcal{E}}\}$.

Corollary 4.1.1 A subset of X is a basis of G iff it is a basis of every P in $\mathcal{P}$.

Theorem 4.2 The G-rank of a subset of X is its minimum P-rank over all P in $\mathcal{P}$.

The proof uses Cor. 2.2.3, Thm. 3.3, and the well-known formula for the rank function of a transversal pregeometry.

Theorem 4.3 Every G-flat is a P-flat for some P in $\mathcal{P}$.

The proof is based on Prop. 2.3.

The dual form of Theorem 4.1 is (Bondy and Welsh [1], Thm.4)

Theorem 4.4 Let H be a pregeometry of rank n on X. Then there exists a family $\mathcal{T}$ of transversal pregeometries of rank n on X such that a subset spans H iff it spans every T in $\mathcal{T}$.

Let (3.2) be an elementary factorization of the closure map of $G = H^*$. The family $\mathcal{T} = \{T(\underline{E}) | \underline{E} \in \underline{\mathcal{E}}\}$ then satisfies the theorem.

Corollary 4.4.1 A subset of X is a basis of H iff it is a basis of every T in $\mathcal{T}$.

Theorem 4.5 The H-rank of a subset of X is its minimum T-rank over all T in $\mathcal{T}$.

This is a direction consequence of Thm. 4.2.

Theorem 4.6 Every H-flat is a T-flat for some T in $\mathfrak{T}$.

The proof is based on Prop. 2.3 and Theorem 4.3.

REFERENCES

1. J. A. Bondy and D. J. A. Welsh, Some Results on Transversal Matroids and Constructions for Identically Self-Dual Matroids. Quart. J. Math. Oxford (2), 22 (1971), 435-451.

2. Terrence J. Brown, Transversal Theory and F-Products. Preprint, University of Missouri at Kansas City 64110.

3. Thomas H. Brylawski, The Tutte-Grothendieck Ring. Ph.D. dissertation, Dartmouth College, Hanover, N. H., 1970.

4. Henry H. Crapo, Single-Element Extensions of Matroids. J. Res. Nat. Bur. Standards Sect. B 69B (1965) 55-65.

5. Henry H. Crapo and Gian-Carlo Rota, On the Foundations of Combinatorial Theory: Combinatorial Geometries (preliminary edition), M.I.T. Press, Cambridge, Mass., 1970.

6. Jack Edmonds and D.R. Fulkerson, Transversals and Matroid Partition. J. Res. Nat. Bur. Standards Sect. B 69B (1965) 147-153.

7. D. A. Higgs, Strong Maps of Geometries. J. Comb. Th. 5 (1968) 185-191.

8. L. Mirsky, Transversal theory. Academic Press, New York, 1971 (Vol. 75 in Mathematics in Science and Engineering.)

SOME PROBLEMS IN GRAPH THEORY

P. Erdös, Hungarian Academy of Science

In this short note I discuss two problems which we discussed during the hypergraph meeting.

I. The following conjecture is attributed to I. Zarins by Bondy and Chvátal: Let k be an integer and $n \geq n_0(k)$. Then if $G(n)$ ($G(n)$ is a graph of n vertices) is Hamiltonian and does not have k pairwise independent vertices (i.e. the complementary graph does not contain a complete subgraph of k vertices) then $G(n)$ is <u>pancyclic</u>, in other words it contains a circuit C_r for every $3 \leq r \leq n$. Bondy and Chvátal inform me that Zarins proved this for $k \leq 4$.

Using a method of Bondy, Chvátal and myself [1], I prove this conjecture. In fact I prove the following

<u>Theorem</u>. <u>Let</u> $n > 4k^4$ <u>and</u> $G(n)$ <u>a Hamiltonian graph which does not contain a set of</u> k <u>pairwise independent vertices</u>. <u>Then</u> $G(n)$ <u>is pancyclic</u>.

A theorem of Bondy and myself states that if $n \geq (k-1)(m-1)+1$ and $n \geq k^2-2$ then $G(n)$ either contains a C_m or k vertices which are mutually independent. Thus if $m \leq \frac{n}{k}$ our $G(n)$ must contain a C_m, and since by assumption $G(n)$ contains a C_n (i.e. it is Hamiltonian) it suffices to prove that $G(n)$ contains a C_m for $\frac{n}{k} < m < n$.

To complete our proof we only have to show that if

$$\frac{n}{k} < m \leq n \tag{1}$$

and if

$G(n)$ contains a C_m it also contains a C_{m-1}. Let $x_1, \ldots, x_m$ be the vertices of our C_m; (x_i, x_{i+1}), $(x_{m+1} = x_1)$ $i = 1, \ldots, m$ are its edges. By (1) and $n > 4k^4$ we can assume that $m > 4k^3$. First of all observe that the set of k vertices $x_i, x_{i+2}, \ldots, x_{i+2(k-1)}$ can not all be independent (since otherwise $G(n)$ would contain a set of k pairwise independent vertices). Thus the graph $G(x_1, \ldots, x_m)$ spanned by the vertices $x_1, \ldots, x_m$ contains at least $2k^2$

(2) (x_{i_r}, x_{j_r}) , $2k > j_r - i_r \geq 2$, $1 \leq i_1 < j_1 < \ldots < i_s < j_s \leq n$, $s \geq 2k^2$.

In fact we can assume $j_r - i_r > 2$ for if $j_r - i_r = 2$, we already have our C_{m-1} (our C_m without the vertex $x_{i_r+1} = x_{j_r-1}$) . We can also assume that no two vertices

(3) (x_u, x_v) , $i_r < u < v < j_r$, $v - u > 1$

are joined (otherwise we could replace (x_{i_r}, x_{j_r}) by (x_u, x_v)) . An edge (x_{i_r}, x_{j_r}) is called good if the valencies (in $G(x_1, \ldots, x_m)$) of every x_u , $i_r < u < j_r$ is $\geq k+2$. Observe that there must be at least one good edge for otherwise we would have at least $2k^2$ vertices of valency $\leq k+1$ and thus easily get a set of k pairwise independent vertices. Assume now that the edge (x_{i_r}, x_{j_r}) is good. Renumber for convenience of notation the vertices of C_m so that

$$y_1 = x_{i_r+1},\ y_2 = x_{i_r+2}, \ldots, y_t = x_{j_r}, \ldots, y_m = x_{i_r},\ t = j_r - i_r .$$

Let C_ℓ, $\ell < m$, be the largest circuit in $G(y_1, \ldots, y_m)$ which contains all the y_u , $t \leq u \leq m$ and perhaps some of the y_u , $1 \leq u < t$. If $\ell = m-1$ our proof is finished (we have our C_{m-1}) . Thus assume $\ell < m-1$, and this assumption will lead to a contradiction. Denote by $v(y_i)$ the valency of y_i . Since the edge $(y_m, y_t) = (x_{i_r}, y_{j_r})$ was good we have

(4) $v(y_u) \geq k+2,\ 1 \leq u < t$.

Let y_u be a vertex of C_m which is not a vertex of C_ℓ .
By (3) y_u is joined only to y_{u-1} and y_{u+1} in $G(y_1, \ldots, y_{t-1})$. Thus by (4) y_u is joined to at least k vertices of C_ℓ . Renumber again the vertices of C_ℓ: $z_1, z_2, \ldots, z_\ell, z_{\ell+1} = z_1$, so that (z_i, z_{i+1}) are the edges of C_ℓ . Let y_u be joined to $z_{i_1}, \ldots, z_{i_k}$. The vertices z_{i_r+1} , $1 \leq r \leq k$ must be independent for if any two of them are joined we get a $C_{\ell+1}$ containing $z_1, \ldots, z_\ell$ and y_u . But our graph can not have k pairwise independent vertices and this contradiction proves $\ell = m-1$ and our theorem.

The inequality $n > 4k^4$ is undoubtedly not best possible; perhaps the result holds for $n > C_1 k^2$ if C_1 is sufficiently large. A simple example shows that it certainly fails for $n < k^2/4$.

II. I state a few results on random graphs; proofs and more detailed statements of the results will be published later.

Denote by $G_r(n; k)$ an r - graph of n vertices and k edges (i.e. k r - tuples). It can be shown by probabilistic methods that for every fixed $\alpha > 0$ if $n \to \infty$, there is an r - graph

$$G_r(n; [\alpha n^r])$$

so that for every $m > c(\alpha, \varepsilon)(\log n)^{1/r-1}$ every spanned subgraph $G_r(m)$ of our $G_r(n)$ having m vertices has more than $(\alpha - \varepsilon)m^r$ and **fewer than** $(\alpha + \varepsilon)m^r$ r - tuples. In other words the distribution of the r - tuples is very uniform. It can also be shown that the above result is best possible, in other words it fails if $c(\alpha, \varepsilon)$ is sufficiently small. This and other related results will be discussed in our forthcoming book with Y. Spencer on applications of probability methods to combinatorial analysis.

For simplicity let us restrict ourselves to ordinary graphs, i.e. to $r = 2$ (this is not essential). Consider graphs $G(n; [n^{1+\alpha}])$ where $0 < \alpha < 1$ and in fact let $\alpha = \frac{1}{2}$. Denote by $f(G(m))$ the maximum number of edges of a subgraph of m vertices of our $G(n; [n^{3/2}])$.

Theorem. There is a $G(n; [n^{\frac{3}{2}}])$ so that for every $m < n^{\frac{1}{2}-\varepsilon}$

$$f(G(m)) < \frac{2}{\varepsilon} m$$

but for every $G(n; [n^{3/2}])$ and for some $m < n^{3/2-\varepsilon}$

$$f(G(m)) > \frac{m}{2\varepsilon}.$$

Further there is a $G(n; [n^{3/2}])$ so that

$$f(G[n^{1/2}]) < \frac{c_1 \log n}{\log \log n} n^{\frac{1}{2}}$$

but for every $G(n; [n^{3/2}])$

$$f(G[n^{\frac{1}{2}}]) > \frac{c_2 \log n}{\log \log n} n^{\frac{1}{2}}.$$

The method of proof is similar to those used by Spencer and myself [2].

REFERENCES

1. Bondy, A., and Erdös, P., Ramsey numbers for cycles in graphs, will appear in the Journal of Combinatorial Theory.

2. Chvátal, V., and Erdös, P., A note on Hamiltonian circuits, Discrete Math. 2(1972), 111-113.

3. Erdös, P. and Spencer, I., Imbalances in k - colorations, Networks, 1, (1972), 379-385.

ASPECTS OF THE THEORY OF HYPERMATROIDS

Thorkell Helgason, University of Iceland

Abstract.

The concept of a matroid is generalized by imitating the generalization of a graph to a hypergraph.

Two basic examples of such hypermatroids are given. The first type models the coloring problem for hypergraphs whereas the second one is related to coverings of hypergraphs. This second example, covering hypermatroids, is the proper extension of trivial matroids. Covering hypermatroids are characterized by a generalized semimodularity condition.

It is shown how all hypermatroids can be naturally constructed from suitable matroids.

A Poincaré polynomial is defined for hypermatroids. The coloring polynomial for hypergraphs is seen to be a special case. For covering hypermatroids the Poincaré polynomial yields a polynomial identity which gives an enumerative relationship between the covering sets and the transversal sets of a hypergraph.

1. Definitions and Examples

1.1 We recall (see e.g. [4]) that a matroid on a finite set S can be characterized by a Whitney rank function. This is a function r, which has the subsets of S as its arguments, takes non-negative integral values and is

(i) normalized: $r(\Phi) = 0$, and $r(\{p\}) \leq 1$ for all $p \in S$.

(ii) monotone increasing: $r(A) \leq r(B)$ if $A \subseteq B \subseteq S$.

(iii) semimodular: $d(A,B) := r(A) + r(B) - r(A \cup B) - r(A \cap B)$ is non-negative for all subsets $A,B \subseteq S$.

We denote the matroid by $[S,r]$.

1.2 Example.

Let S be a finite set. The trivial matroid on S is the pair $[S,r]$ with $r(A) := |A|$, where $A \subseteq S$. Evidently trivial matroids can be characterized by the condition

$d(A,B) = 0$ for all $A,B \subseteq S$.

1.3 Example.

Let $(V,\mathcal{E})$ be a graph with vertex set V and family of edges $\mathcal{E}$. A rank function r may be defined on $\mathcal{E}$ by putting $r(\mathcal{A}) :=$ (number of vertices adjacent to $\mathcal{A}$) - (number of connected components of $\mathcal{A}$) where $\mathcal{A} \subseteq \mathcal{E}$.

Crapo and Rota call the matroid $[\mathcal{E},r]$ the bond geometry (matroid) of the graph (see [4]) whereas in their expository paper [8] Harary and Welsh call it the cyclic matroid. We propose the name chromatic matroid, because the coloring problem of the graph can be formulated in terms of this matroid.

We recall that the chromatic polynomial of the graph is identical to the characteristic polynomial of the "chromatic" matroid. This is the polynomial

$$\chi(\lambda) = \sum_{\mathcal{A} \in \mathcal{B}} (-1)^{|\mathcal{A}|} \lambda^{r(\mathcal{B}) - r(\mathcal{A})}$$

For details; [10].

1.4 Interest in the (weak) coloring problem for hypergraphs leads to the question, whether the chromatic polynomial for hypergraphs can be expressed in a similar way. By a hypergraph (see [1]) we understand a pair $H = (V,\mathcal{B})$ where V is a finite set whose elements are called the vertices of the hypergraph, and where $\mathcal{B}$ is a finite family of subsets of V which may be repeated and are called the hyperedges of the hypergraph. The empty set may be taken as a hyperedge. We will always assume that every vertex is a member of some hyperedge.

To deal with the above question on the chromatic polynomial for hypergraphs we introduce the concepts of a hyperrank function and of a hypermatroid.

1.5 Let S be a finite set. A hyperrank function on S is a function r, having as arguments the subsets of S, taking non-negative integers as values, and being monotone increasing, semimodular and

(i) seminormalized: $r(\Phi) = 0$ and $r(S) > 0$.

The pair $[S,r]$ is called a hypermatroid. Every matroid is trivially a hypermatroid. There is a wide selection of hypermatroids. In section 2 we will show how they can all be constructed. We shall otherwise, only be concerned with those hypermatroids which are associated with hypergraphs.

1.6 Example

Let H = (V, $\mathcal{R}$) be a hypergraph. For a family $\mathcal{A}$ of hyperedges the set $\cup\mathcal{A}$ denotes the set of vertices adjacent to the hyperedges in $\mathcal{A}$ and $|\cup\mathcal{A}|$ denotes the cardinality of this set. The function r, defined by

$$r(\mathcal{A}) := |\cup\mathcal{A}| \text{ for } \mathcal{A} \subseteq \mathcal{R}.$$

is a hyperrank function on $\mathcal{R}$. We will call the hypermatroid $[\mathcal{R}, r]$ the covering hypermatroid of the hypergraph. Reasons for using this name will be found in 3.12. Covering hypermatroids are the simplest hypermatroids that we can associate with a hypergraph. Furthermore we believe they constitute the proper generalization of the trivial matroids. The characterization of the trivial matroids is quite easy whereas more effort is needed to characterize the covering hypermatroids.

1.7 Example.

Again let H = (V, $\mathcal{R}$) be a hypergraph, A hyperrank function r may now be defined on $\mathcal{R}$ by putting

$$r(\mathcal{A}) := |\cup\mathcal{A}| - (\text{number of connected components of the hypergraph } (\cup\mathcal{A}, \mathcal{A}))$$

for a subfamily $\mathcal{A}$ of $\mathcal{R}$. We call this hypermatroid the chromatic hypermatroid of the hypergraph. As will be seen in 3.14, this hypermatroid contains the information needed for the (weak) coloring problem of the hypergraph.

1.8 Now we endeavour a characterization of covering hypermatroids, but first we make precise the notion of isomorphism, and introduce some notation.

1.9 Two hypermatroids $[S_1, r_1]$ and $[S_2, r_2]$ are called isomorphic if there exists an injective map ϕ of S_1 onto S_2 with

$$r_1(A) = r_2(\phi(A)) \text{ if } A \subseteq S_1.$$

1.10 If $[S,r]$ is a hypermatroid, then we define

$$c(S) := \text{the number of elements } p \in S \text{ such that } 0 < r(\{p\}) < r(S)$$

and

$$D(A,B) := \sum_{C \subseteq S - A \cap B} (-1)^{|C|} d(A \cup C, B \cup C)$$

where $A,B \subseteq S$. See 1.1 for the definition of d.

1.11 <u>Theorem</u>. <u>A hypermatroid</u> $[S,r]$ <u>is isomorphic to the covering hypermatroid of some hypergraph if and only if</u>

$$D(A,B) \geq 0 \quad \underline{\text{for all subsets}} \quad A,B \subseteq S$$

Proof: We first treat the case $c(S) = 0$ and then proceed by induction on the number $r(S)$.

Suppose the hypermatroid $[S,r]$ is such that $c(S) = 0$. We will show that such a hypermatroid is always isomorphic to a covering hypermatroid. Let V be a vertex set with r(S) vertices and let $\mathcal{E} = (E_p : p \in S)$ where $E_p = \phi$ if $r(\{p\}) = 0$ and $E_p = V$ if $r(\{p\}) = r(S)$. Now the hypermatroid $[S,r]$ is isomorphic to the covering hypermatroid of $(V,\mathcal{E})$.

Next we want to prove that $D(A,B) = 0$ for all subsets $A,B \subseteq S$ given that $c(S) = 0$.

Let $F = \{p \in S: r(\{p\}) = r(S)\}$ and take $M,N \subseteq S$. Evidently $r(M) = r(S)$ if $M \cap F \neq \phi$ and $r(M) = 0$ if $M \cap F = \phi$. Therefore

$$d(M,N) = r(M) + r(N) - r(M \cup N) - r(M \cap N)$$

is zero unless $M \cap F \neq \phi$, $N \cap F \neq \phi$ and $M \cap N \cap F = \phi$ in which case $d(M,N) = r(S)$.

Hence if $A,B,C \subseteq S$ then $d(A \cup C, B \cup C) = 0$ unless $A \cap F \neq \phi$, $B \cap F \neq \phi$, $A \cap B \cap F = \phi$ and $C \cap F = \phi$ in which case $d(A \cup C, B \cup C) = r(S)$. Therefore $D(A,B) = 0$ unless $A \cap F \neq \phi$, $B \cap F \neq \phi$ and $A \cap B \cap F = \phi$ in which case we have

$$D(A,B) = \sum_{C \subseteq S - A \cap B} (-1)^{|C|} d(A \cup C, B \cup C) = r(S) \sum_{C \subseteq S - (A \cap B) \cup F} (-1)^{|C|}$$

which equals 0 if $(A \cap B) \cup F \neq S$ but equals $r(S)$ if $(A \cap B) \cup F = S$. In any case $D(A,B) \geq 0$.

The proof of theorem 1.11 now follows by induction on the number $r(S)$. If $r(S) = 1$ then $c(S) = 0$ and the above applies. Suppose we have proved the theorem for all hypermatroids of rank less than m. Now suppose $r(S) = m$ and $c(S) \neq 0$. Thus there is a $p \in S$ with $0 < r(\{p\}) < r(S)$. We fix this p and define new functions r_p and r^p by putting

$$r_p(A) = r(A \cup \{p\}) - r(\{p\}) \text{ and}$$

$$r^p(A) = r(A) + r(\{p\}) - r(A \cup \{p\})$$

for every $A \subseteq S$. Obviously $r = r_p + r^p$ and hence

$d = d_p + d^p$ and $D = D_p + D^p$.

Suppose $[S,r]$ satisfies the condition $D \geq 0$. We will now show that $D_p \geq 0$ and $D^p \geq 0$. Let $A,B \subseteq S$ and first consider the case $p \in A \cap B$. Then $d_p(A \cup C, B \cup C) = d(A \cup C, B \cup C)$ for all $C \subseteq S$ and hence $D_p(A,B) = D(A,B) \geq 0$. But then $D^p(A,B) = 0$. Secondly in the case $p \notin A \cap B$ we have

$$d_p(A \cup C, B \cup C) = d(A \cup C \cup \{p\}, B \cup C \cup \{p\}) \text{ and}$$

$$D_p(A,B) = \sum_{\substack{C \subseteq S - A \cap B \\ p \notin C}} (-1)^{|C|} d(A \cup C \cup \{p\}, B \cup C \cup \{p\})$$

$$+ \sum_{\substack{C \subseteq S - A \cap B \\ p \in C}} (-1)^{|C|} d(A \cup C \cup \{p\}, B \cup C \cup \{p\})$$

$$= \sum_{\substack{C \subseteq S - A \cap B \\ p \notin C}} (-1)^{|C|} d(A \cup C \cup \{p\}, B \cup C \cup \{p\})$$

$$+ \sum_{\substack{C \subseteq S - A \cap B \\ p \notin C}} (-1)^{|C|+1} d(A \cup C \cup \{p\}, B \cup C \cup \{p\}) = 0$$

Therefore $D^p(A,B) = D(A,B) \geq 0$ and $D_p(A,B) = 0$. Next we want to check that r_p and r^p satisfy the axioms of a hyperrank function on S:

(i) seminormalized: $r_p(\phi) = r(\{p\}) - r(\{p\}) = 0$, $r_p(S) = r(S) - r(\{p\}) > 0$ and $r^p(\phi) = r(\phi) + r(\{p\}) - r(\{p\}) = 0$, $r^p(S) = r(S) + r(\{p\}) - r(S) = r(\{p\}) > 0$

(ii) monotone increasing: Let $A \subseteq S$ and $q \in S$. Then

$$r_p(A \cup \{q\}) = r(A \cup \{q\} \cup \{p\}) - r(\{p\})$$
$$\geq r(A \cup \{p\}) - r(\{p\}) = r_p(A) \text{ and}$$
$$r^p(A \cup \{q\}) = r(A \cup \{q\}) + r(\{p\}) - r(A \cup \{q\} \cup \{p\})$$
$$\geq r((A \cup \{q\}) \cap (A \cup \{p\})) + r(\{p\}) - r(A \cup \{p\})$$

$\geq r(A) + r(\{p\}) - r(A\cup\{p\}) = r^p(A)$.

(iii) semimodular: Since $D_p \geq 0$ and $D^p \geq 0$ semimodularity follows from Lemma 1.12.

We have now showed that $[S,r_p]$ and $[S,r^p]$ are hypermatroids and that $D_p \geq 0$ and $D^p \geq 0$. Since $r_p(S) = r(S) - r(\{p\}) < r(S)$ and $r^p(S) = r(\{p\}) < r(S)$ induction hypothesis applies. Thus there exist two hypergraphs $H_p = (V_p, \mathcal{E}_p)$ and $H^p = (V^p, \mathcal{E}^p)$ and bijective mappings.

$$\phi_p : S \to \mathcal{E}_p \quad \text{and}$$

$$\phi^p : S \to \mathcal{E}^p$$

which give an isomorphism from the hypermatroid $[S,r_p]$, respectively $[S,r_p]$, to the covering hypermatroid of the hypergraph H_p, respectively H^p. Now let V be the disjoint union of the sets V_p and V^p, $V := V_p \dot{\cup} V^p$. Put $\Phi(q) = \Phi_p(q) \cup \phi^p(q)$ for $q \in S$ and let $\mathcal{E} = (\Phi(q) : q \in S)$. Now

$$\Phi : S \to \mathcal{E}$$

and for $A \subseteq S$

$$r(A) = r_p(A) + r^p(A) = |\bigcup_{q\in A} \Phi_p(q)| + |\bigcup_{q\in A} \phi^p(q)|$$

$$= |\bigcup_{q\in A} \Phi(q)|$$

which shows that $[S,r]$ is isomorphic to the covering hypermatroid of $H = (V, \mathcal{E})$.

Conversely suppose that the map

$$\Phi : S \to \mathcal{E}$$

gives an isomorphism from the hypermatroid $[S,r]$ to the covering hypermatroid of the hypergraph $H = (V, \mathcal{E})$. We want to show that $D \geq 0$. Define $H_p = (V_p, \mathcal{E}_p)$ where $V_p = V - \Phi(p)$ and $\mathcal{E}_p = (E - \Phi(p) : E \in \mathcal{E})$ and $H^p = (V^p, \mathcal{E}^p)$ where $V^p = \Phi(p)$ and $\mathcal{E}^p = (E \cap \Phi(p) : E \in \mathcal{E})$

Furthermore define

$$\Phi_p : S \to \mathcal{E}_p \quad \text{and}$$

$$\Phi^p : S \to \mathcal{E}^p$$

by putting $\Phi_p(q) = \Phi(q) - \Phi(p)$ and $\Phi^p(q) = \Phi(q) \cap \Phi(p)$. Clearly the map Φ_p (respectively Φ^p) gives an isomorphism from $[S, r_p]$ (respectively $[S, r^p]$) to the covering hypermatroid of H_p (respectively H^p). In particular this shows that $[S, r_p]$ and $[S, r^p]$ are hypermatroids and using the induction hypothesis we have $D_p \geq 0$ and $D^p \geq 0$. But then $D = D_p + D^p \geq 0$.

1.12 <u>Lemma</u>. <u>For a function</u> $d(M,N)$ <u>having as arguments subsets</u> M <u>and</u> N <u>of</u> S <u>the condition</u>

$$\text{(i)} \quad \sum_{C \subseteq S - A \cap B} (-1)^{|C|} d(A \cup C, B \cup C) \geq 0 \quad \underline{\text{for all}} \quad A, B \subseteq S$$

<u>implies the condition</u>

$$\text{(ii)} \quad d(A,B) \geq 0 \quad \underline{\text{for all}} \quad A, B \subseteq S$$

Proof. Assume the first condition holds. Instead of proving the validity of the second inequality we will prove the more general condition

$$\text{(iii)} \quad \sum_{C \subseteq M} (-1)^{|C|} d(A \cup C, B \cup C) \geq 0 \quad \text{for all} \quad A, B \subseteq S$$

By taking $M = \phi$ this shows that $d(A,B) \geq 0$. Now let us prove this last condition by induction on the number of elements in the set $(S - A \cap B) - M$. First if $S - A \cap B = M$ inequality (iii) is the same as inequality (i). So suppose there exists a $q \in (S - A \cap B) - M$. Then

$$\sum_{C \subseteq M \cup \{q\}} (-1)^{|C|} d(A \cup C, B \cup C) =$$

$$\sum_{C \subseteq M} (-1)^{|C|} d(A \cup C, B \cup C) - \sum_{C \subseteq M} (-1)^{|C|} d(A \cup C \cup \{q\}, B \cup C \cup \{q\})$$

or

$$\sum_{C \subseteq M} (-1)^{|C|} d(A \cup C, B \cup C) =$$

$$\sum_{C \subseteq M \cup \{q\}} (-1)^{|C|} d(A \cup C, B \cup C) + \sum_{C \subseteq M} (-1)^{|C|} d(A \cup \{q\} \cup C, B \cup \{q\} \cup C)$$

$$\geq 0 + 0 = 0 \text{ since } |(S - A \cap B) - M \cup \{q\}|$$

and $|(S - (A \cup \{q\}) \cap (B \cup \{q\})) - M|$ are less than $|(S - A \cap B) - M|$ so the induction hypothesis can be applied.

1.13 We remark that there are hypermatroids which are not isomorphic to covering hypermatroids, e.g. the hypermatroid $[S,r]$ where $S = \{a,b,c\}$ and $r(\{x\}) = 1$ for $x \in S$ and $r(\{x,y\}) = 2$ for $x,y \in S, x \neq y$ and $r(S) = 2$. Now $D(\{a\},\{b\}) < 0$.

We also wish to remark that condition $D \geq 0$ in theorem 1.11 can not be changed to $D = 0$ as can be seen by considering the covering hypermatroid of the hypergraph $H = (V, \mathcal{E})$ where $V = \{x,y,z,w\}$ and $\mathcal{E} = (\{x\},\{x,y,z\},\{x,y,w\})$ since $D((\{x\},\{x,y,z\}),(\{x\},\{x,y,w\})) > 0$.

2. A Characterization of Hypermatroids.

2.1 Let $[S',r']$ be any matroid and $\sigma: S \to P(S')$ a map from a finite set S into the power set of S'. Then the function r defined by

$$r(A) := r'(\bigcup \sigma(A)) \quad \text{for} \quad A \subseteq S$$

is a hyperrank function on S.

We note that both covering and chromatic hypermatroids are of this type. Indeed every hypermatroid arises in this way as will be shown.

2.2 Let $[S_1,r_1]$ and $[S_2,r_2]$ be two hypermatroids. A map from the first to the second, denoted by

$$f : [S_1,r_1] \to [S_2,r_2]$$

is a mapping

$$f : S_1 \to S_2 \cup \{\phi\}$$

satisfying $r_2(f(A)) \leq r_1(A)$ for all $A \subseteq S_1$.

2.3 Given a matroid $[S',r']$ let $[P(S'),P(r')]$ be the hypermatroid given by

$$P(r')(A) = r'(\bigcup A) \quad \text{for} \quad A \subseteq P(S').$$

2.4 A <u>representation</u> of the hypermatroid $[S,r]$ by the matroid $[S',r']$ is a map

$$\sigma : [S,r] \to [P(S'),P(r')]$$

satisfying

$r(A) = r'(\bigcup\sigma(A))$ for $A \subseteq S$.

The representation is _faithful_ if σ is one-to-one and

$\bigcup\sigma(S) = S'$.

2.5 _Proposition. Let_ $[S,r]$ _be a hypermatroid. There exists a matroid_ $[S_D,r_D]$ _together with a faithful representation_

$$\sigma : [S,r] \to [P(S_D),P(r_D)]$$

Proof. The matroid $[S_D,r_D]$ is constructed using the Dilworth completion of a semimodular function (see [5] and also [6]).

For each point $p \in S$ choose a set with $r(\{p\})$ elements and denote it by p_D. Then define S_D to be the disjoint union of these sets, i.e.

$$S_D = \dot{\bigcup_{p\in S}} p_D.$$

For $B \subseteq S_D$ let $S(B) = \{p \in S : p_D \cap B \neq \emptyset\}$.
For any $B_1, B_2 \subseteq S_D$ we have

$$S(B_1 \cup B_2) = S(B_1) \cup S(B_2)$$

$$S(B_1 \cap B_2) \subseteq S(B_1) \cap S(B_2) \quad \text{and}$$

$$S(B_1) \subseteq S(B_2) \quad \text{if} \quad B_1 \subseteq B_2.$$

Now we call a subset B of S_D _independent_ if $r(S(B')) \geq |B'|$ for all subsets B' of B.

2.6 Lemma. The family of minimal dependent subsets of S_D satisfies the circuit elimination condition for matroids (see [4]).

Proof. Let B be a minimal dependent subsets of S_D. Then $r(S(B)) \leq |B| - 1$ but $r(S(B')) \geq |B'|$ for $B' \subsetneq B$.

Assume that B_1, B_2 are distinct minimal subsets and $x \in B_1 \cap B_2$. Now

$$r(S(B_1 \cup B_2 - \{x\})) \leq r(S(B_1 \cup B_2))$$

$$= r(S(B_1) \cup S(B_2)) \leq r(S(B_1)) + r(S(B_2)) - r(S(B_1) \cap S(B_2))$$

$$\leq r(S(B_1)) + r(S(B_2)) - r(S(B_1 \cap B_2))$$

$$\leq |B_1| - 1 + |B_2| - 1 - |B_1 \cap B_2|$$

$$= |B_1 \cup B_2| - 2 < |B_1 \cup B_2 - \{x\}|.$$ Hence

$B_1 \cup B_2 - \{x\}$ is dependent and contains a minimal dependent set. So the family of minimal dependent subsets of S_D satisfies the circuit elimination condition.

2.7 We now continue with the proof of proposition 2.5.

Let r_D be the rank function of the matroid on S_D whose existence follows from the Lemma 2.6. We recall that

$$r_D(B) = |C|$$

where C is any maximal independent subset of B.

We want to show that the map

$$\sigma : S \to P(S_D) \quad \text{defined by} \quad p \mapsto p_D$$

is a faithful representation

$$\sigma : [S,r] \to [P(S_D),P(r_D)]$$

Since σ is clearly one-to-one and $\sigma(S) = S_D$ it remains to be shown that $r(A) = r_D(\bigcup\sigma(A))$ for all $A \subseteq S$. Let $B \subseteq S_D$ be a maximal independent subset of $\bigcup\sigma(A)$: Then $S(B) \subseteq A$ and therefore $r(A) \geq r(S(B)) \geq |B| = r_D(\bigcup\sigma(A))$. Conversely we use induction on $|A|$ to show that $r_D(\bigcup\sigma(A)) \geq r(A)$.

For $A = \Phi$ this inequality is obvious. Let $p\varepsilon S-A$ and suppose we have proved that $r_D(\bigcup\sigma(A)) \geq r(A)$. Let $B \subseteq S_D$ be a maximal independent subset of $\bigcup\sigma(A)$. Let $b \subseteq p_D$ be a subset with $(r(A\cup\{p\}) - r(A))$ elements. Note that $|b| = r(A\cup\{p\}) - r(A) \leq r(\{p\}) = |p_D|$. We claim that $B\cup b$ is independent in $[S_D,r_D]$. Let $B' \subseteq B$ and $b' \subseteq b$, $b' \neq \Phi$. Then $S(B') \subseteq A$ and $S(B'\cup b') = S(B')\cup\{p\}$. Therefore

$$r(S(B'\cup b') + r(A) \geq r(S(B'\cup b')\cup A) + r(S(B'\cup b')\cap A)$$

$$= r(A\cup\{p\}) + r(S(B')).$$ Hence

$$r(S(B'\cup b')) \geq r(S(B')) + r(A\cup\{p\}) - r(A)$$

$$\geq |B'| + |b'| = |B'\cup b'|.$$ Together with

$r(S(B')) \geq |B'|$ this shows that $B\cup b$ is independent.

But then $r(A\cup\{p\}) = r(A) + r(A\cup\{p\}) - r(A)$

$$\leq r_D(\bigcup\sigma(A)) + |b| = |B| + |b| = |B\cup b| \leq r_D(\bigcup\sigma(A\cup\{p\}))$$

which completes the induction.

2.8 A more detailed formulation of proposition 2.5 can be given using the terminology of category theory (see [9]).

Let $f,g : [S_1,r_1] \rightarrow [S_2,r_2]$ be two maps (in the sense of 2.2). We say that f and g are equivalent if $r_2(f(p)) = r_2(f(p) \cup g(p)) = r_2(g(p))$ for all $p \in S_1$. Obviously this is an equivalence relation on the set of maps. Let us call such an equivalence class of maps a mapclass. Hypermatroids together with their mapclasses form a category. Matroids form a full subcategory.

The assignment of the hypermatroid $[P(S),P(r)]$ to the matroid $[S,r]$ easily extends to a functor P from the category of matroids to the category of hypermatroids.

The generalization of proposition 2.5 we are aiming at can now be stated as follows:

2.9 Theorem. The functor P has a left adjoint.

The essence of this theorem is contained in proposition 2.5 so we will leave out a detailed proof of it.

3. The Poincaré Polynomial for Hypermatroids.

3.1 The concept of the Poincaré polynomial for matroids (see [3]) is immediately generalizable to hypermatroids.

Let $[S,r]$ be a hypermatroid. We call the polynomial

$$\tau_{[S,r]}(\lambda,\eta) = \sum_{A \subseteq S} (\eta-1)^{|A|} \lambda^{r(S) - r(A)}$$

the Poincaré polynomial of the hypermatroid $[S,r]$. Then

$$\chi_{[S,r]}(\lambda) = \tau_{[S,r]}(\lambda,0)$$

is called the <u>characteristic polynomial</u> of the hypermatroid $[S,r]$.

3.2 Given a matroid $[S',r']$ and subset $B \subseteq S$ the <u>contraction</u> of $[S',r']$ by B, denoted by $[S',r']/B$, is a matroid on the point set $S'- B$ with the rank function $\bar{r}$ defined by

$$\bar{r}(A) = r'(A \cup B) - r'(B) \quad \text{for} \quad A \subseteq S - B.$$

Our main result on the Poincaré polynomial is contained in the following theorem.

3.3 <u>Theorem.</u> <u>Let</u>

$$\sigma : [S,r] \to [P(S'),P(r')]$$

<u>be a representation of the hypermatroid</u> $[S,r]$ <u>by the matroid</u> $[S',r']$ <u>and assume</u> $r(S) = r'(S')$.

<u>Then</u>

(i) $$\tau_{[S,r]}(\lambda,\eta) = \sum_{B \subseteq S} \eta^{|\sigma^{\Delta}(B)|} \chi_{[S',r']/B}(\lambda)$$

<u>where</u> $\sigma^{\Delta}(B) = \{p \in S : \sigma(p) \subseteq B\}$

<u>Furthermore</u>

(ii) $$\tau_{[S,r]}(\lambda,\eta) = \sum_{B \in L_{r'}(S')} \eta^{|\sigma^{\Delta}(B)|} \chi_{[S',r']/B}(\lambda)$$

<u>where</u> $L_{r'}(S')$ <u>is the lattice of closed subsets of the matroid</u> $[S',r']$, <u>i.e. maximal subsets of</u> S <u>with a given rank</u>.

3.4 In the proof of this theorem the following lemma of H. Crapo (see [2]) is used.

3.5 Lemma. *If $\alpha : P \to L$ is a sup-homomorphism from a finite lattice P into a finite lattice L, if f is a function on P and g a function on L, both into some ring, then*

$$\sum_{x\in P} f(x)g(\alpha(x)) = \sum_{y\in L} Sf(\alpha^{\Delta}(y))\ Eg(y)$$

where

$$\alpha^{\Delta}(y) = \sup\ \{x\in P : \alpha(x) \leq y\}$$

$$Sf(x) = \sum_{\substack{z\in P\\ z\leq x}} f(z)$$

$$Eg(y) = \sum_{\substack{z\in L\\ z\geq y}} \mu(y,z)\ g(z),$$ μ *the Möbius function on* L.

3.6 Now let us prove the theorem using the lemma. To prove the identity (i) of the theorem we take α in the lemma to be the map from P(S) into P(S) defined by $A \mapsto \sigma(A)$ and to prove (ii) take α to be the map from P(S) into $L_{r'}(S')$ defined by $A \mapsto \overline{\sigma(A)}$ where $\overline{\sigma(A)}$ is the closed hull of $\sigma(A)$.

Let

$$f(A) = (\eta-1)^{|A|} \quad \text{and}$$

$$g(B) = \lambda^{r'(S') - r'(B)}$$

Then

$$Sf(A) = \sum_{A' \subseteq A} (\eta-1)^{|A'|} = \eta^{|A|} \quad \text{and}$$

for the first choice of α

$$Eg(B) = \sum_{B' \subseteq B} (-1)^{|B'-B|} \lambda^{r'(S') - r'(B')}$$

$$= \sum_{B' \subseteq S'-B} (-1)^{|B'|} \lambda^{(r'(S') - r'(B)) - (r'(B' \cup B) - r'(B))}$$

$$= \chi_{[S',r']/B}(\lambda)$$

The equality $Eg(B) = \chi_{[S',r']/B}(\lambda)$

is proved similarily for the second choice of α.

Furthermore

$$g(\alpha(A)) = \lambda^{r'(S') - r'(\sigma(A))} \quad (\text{or } \lambda^{r'(S') - r'\overline{(\sigma(A))}})$$

$$= \lambda^{r(S) - r(A)}$$

Therefore the lemma states now that

$$\sum_{A \subseteq S} (\eta-1)^{|A|} \lambda^{r(S) - r(A)} = \sum_{B \in L_{r'}(S')} \eta^{|\sigma^{\Delta}(B)|} \chi_{[S',r']/B}(\lambda)$$

3.7 Specialization of theorem (4.1) to our two basic examples of hypermatroids (see 1.6 and 1.7) gives interesting results.

First consider the covering hypermatroid. Let $H = (V,\mathcal{E})$ be a hypergraph with no empty edges, i.e. $E \neq \phi$ for all $E \in \mathcal{E}$.

The inclusion map

$\sigma : \mathcal{E} \to P(V)$ defined by $\sigma(E) = E$

is a representation of the covering hypermatroid $[\mathcal{E},r]$ of the hypergraph H by the trivial matroid $[V,r']$, and $r(\mathcal{E}) = r'(V)$.

Let us denote by $\tau_H(\lambda,\eta)$ the Poincaré polynomial of the covering hypermatroid of H.

Any contraction of a trivial matroid is trivial. Thus the characteristic polynomial of the contraction of the trivial matroid on V by the subset B of V is the polynomial

$$(\lambda-1)^{|V-B|}$$

Hence theorem (3.3) yields the identity

$$\tau_H(\lambda,\eta) = \sum_{B\subseteq V} \eta^{|\sigma^{\Delta}(B)|}(\lambda-1)^{|V-B|}.$$

But

$$\sigma^{\Delta}(B) = (E\varepsilon\mathcal{E} : E\subseteq B) = (E\varepsilon\mathcal{E} : E\cap(V-B) = \Phi).$$

Therefore also

$$\tau_H(\lambda,\eta) = \sum_{B\subseteq V}(\lambda-1)^{|B|}\eta^{|\mathcal{E}| - |(E\varepsilon\mathcal{E} : E\cap B\neq\Phi)|}$$

Now consider the <u>dual</u> hypergraph of H which is defined as the hypergraph $H^* = (V^*,\mathcal{E}^*)$ where

$$V^* = \mathcal{E} \text{ and}$$

$$\mathcal{E}^* = (E_x : x\varepsilon V) \quad \text{with} \quad E_x = (E\varepsilon\mathcal{E} : x\varepsilon E)$$

The right hand side of (i) in 3.3 can thus be written as

$$\sum_{\mathcal{A}^*\subseteq\mathcal{E}^*}(\lambda-1)^{|\mathcal{A}^*|}\eta^{|V^*| - |\cup\mathcal{A}^*|}$$

and since $V^* = \cup\mathcal{E}^*$ we immediately identify this expression as the Poincaré polynomial of the covering hypermatroid of H in the variables η and λ

We have proved

3.8 *Theorem. If* H *is a hypergraph with no empty edges then*

$$\tau_H(\lambda,\eta) = \tau_{H^*}(\eta,\lambda)$$

or equivalently

$$\sum_{\mathcal{A}\subseteq\mathcal{E}} (\eta-1)^{|\mathcal{A}|}\, \lambda^{|V| - |\cup\mathcal{A}|} = \sum_{B\subseteq V} (\lambda-1)^{|B|}\, \eta^{|\mathcal{E}| - |(E\in\mathcal{E}:\ E\cap B \neq \Phi)|}$$

where $H = (V,\mathcal{E})$

3.9 A subset $T \subseteq V$ is a *transversal* set for the hypergraph $H = (V,\mathcal{E})$ if T meets every edge of H.

3.10 *Corollary. If* $H = (V,\mathcal{E})$ *is a hypergraph with no empty edges then the characteristic polynomial of its covering hypermatroid*

$$\chi_H(\lambda) = \sum_{\mathcal{A}\subseteq\mathcal{E}} (-1)^{|\mathcal{A}|}\, \lambda^{|V| - |\cup\mathcal{A}|}$$

is identical to

$$\sum_{T} (\lambda-1)^{|T|}$$

where the sum is taken over all transversal subsets of V.

3.11 The dual concept of a transversal set is a covering set. A subset $\mathcal{C}\subseteq\mathcal{E}$ is a *covering* set for the hypergraph H if $\cup\mathcal{C} = V$.

3.12 Corollary. *If $H = (V,\mathcal{E})$ is a hypergraph with no empty edges then the characteristic polynomial of the covering hypermatroid of H^**

$$\chi_{H^*}(\eta) = \sum_{B \subseteq V} (-1)^{|B|} \eta^{|\mathcal{E}| - |(E\varepsilon\mathcal{E}:\ E \cap B \neq \Phi)|}$$

is identical to

$$\sum_{\mathcal{E}} (\eta-1)^{|\mathcal{E}|}$$

where the sum is taken over all covering subsets of V.

3.13 Secondly let us interpret the main theorem (3.3) for chromatic hypermatroids. Again let $H = (V,\mathcal{E})$ be a hypergraph with no empty edges. Let $K(V)$ be the complete graph on V furnished with the chromatic matroid structure. Now the map

$$\sigma : \mathcal{E} \to K(V) \quad \text{defined by} \quad E \mapsto E_2$$

where $E_2 = \{\{x,y\}\varepsilon K(V) : x,y\varepsilon E,\ x \neq y\}$ is a representing map satisfying the rank equality condition of 3.3.

Let $\tau_H^c(\lambda,\eta)$ denote the Poincaré polynomial of the chromatic hypermatroid on H. We shall use identity (ii) of 3.3 to express this polynomial.

The lattice of closed subsets of $K(V)$ is identical to the lattice of partitions of the set $V,\mathcal{P}(V)$. Given a partition $\pi\varepsilon\mathcal{P}(V)$ then

$$\sigma^{\Delta}(\pi) = (E\varepsilon\mathcal{E} : E \text{ is contained in a block in } \pi)$$

Furthermore the contraction of $K(V)$ by π corresponds to the complete graph on k vertices where k is the number of blocks in π.

It is well known that the characteristic polynomial of the chromatic matroid of a complete graph on k vertices equals

$$(\lambda)_k = (\lambda-1)\cdots(\lambda-k+1).$$

Let $\mathcal{P}_k(V)$ be the subset of $\mathcal{P}(V)$ consisting of all partitions of V into k blocks.

Now we can use identity (ii) of 3.3 to express τ_H^c as follows:

$$\tau_H^c(\lambda,\eta) = \sum_{k=0}^{n} \sum_{\pi\in\mathcal{P}_k(V)} \eta^{|\sigma^\Delta(\pi)|} (\lambda)_k$$

where $n = |V|$. In particular

$$\chi_H^c(\lambda) = \sum_{k=0} a_k^H (\lambda)_k$$

where a_k^H is the number of partitions of V into k blocks none of which contains an hyperedge.

Therefore $\lambda\chi_H^c(\lambda)$ is obviously the (weak) coloring polynomial of the hypergraph H.

We have proved the following theorem which is well known for graphs (see e.g. [10]).

3.14 Theorem. Let $H = (V,\mathcal{E})$ be a hypergraph without empty edges. Then the number of weak λ-colorings of H is $\lambda\,\chi_H^c(\lambda)$ i.e. equals

$$\lambda \sum_{\mathcal{A}\subseteq\mathcal{E}} (-1)^{|\mathcal{A}|} \lambda^{r_c(\mathcal{E}) - r_c(\mathcal{A})}$$

(where r_c is the rank function of the chromatic hypermatroid of H).

REFERENCES

1. Berge, C., Graphes et hypergraphes, Dunod, Paris, 1970.

2. Crapo H., Möbius Inversion in Lattices, Archiv der Math. 19 (1968), 595-607.

3. Crapo H., The Tutte Polynomial, Aequationes Math. 3 (1969), 211-229.

4. Crapo H. and Rota G-C., On the Foundations of Combinatorial Theory: Combinatorial Geometries, M.I.T. Press, Cambridge, 1970.

5. Dilworth, R.P., Dependence Relations in a Semimodular Lattice, Duke Math. J. 11 (1944), 575-587.

6. Edmonds, J., Submodular Functions, Matroids, and Certain Polyhedra, pp. 69-87 in Guy, R. et al. (editors), Combinatorial Structures and their Applications, Gordon and Breach, New York etc., 1970.

7. Harary, F. and Welsh, D., Matroids versus Graphs, pp. 155-170 in Chartrand, G. et al. (editors). The Many Facets of Graph Theory, Springer-Verlag, Berlin etc. 1969.

8. Helgason, T., On hypergraphs and Hypergeometries, thesis Massachusetts Institute of Technology, 1971 supervised by Gian-Carlo Rota.

9. Pareigis, B., Categories and Functors, Academic Press, New York etc., 1970.

10. Rota, G-C., On the Foundations of Combinatorial Theory I The Möbius Functions, Z. Wahrscheinlichkeitsth. und verw. Gebiete, 2 (1964), 340-368.

FACETS OF 1-MATCHING POLYHEDRA

W. Pulleyblank, Jack Edmonds, University of Waterloo

All graphs considered are finite and loopless. Let $G = (V, E, \psi)$ be a graph. V is the node set; E is the edge-set; for every $j \in E$, $\psi(i) \subseteq V$ is the pair of nodes which j meets. For any $S \subseteq V$ we let $\delta(S)$ denote the coboundary of S, that is

$$\delta(S) = \{j \in E: |\psi(j) \cap S| = 1\}.$$

For any $v \in V$ we abbreviate $\delta(\{v\})$ by $\delta(v)$. For any $S \subseteq V$ we let $\gamma(S)$ denote the set of edges having both ends in S, that is

$$\gamma(S) = \{j \in E: \psi(j) \subseteq S\} .$$

For any $S \subseteq V$ we let $G[S]$ denote the subgraph of G induced by S, that is

$$G[S] = (S, \gamma(S), \psi|\gamma(S))$$

where for any $J \subseteq E$ we let $\psi|J$ denote the restriction of ψ to J. If H is any graph we let $V(H)$ and $E(H)$ denote the node set and edge set of H respectively.

$\mathbb{R}$ denotes the set of real numbers. For any finite set X, $\mathbb{R}^X$ denotes the set of all vectors $(x_i \in \mathbb{R}: i \in X)$. For any such X and any $y = (y_i: i \in X) \in \mathbb{R}^X$ and any $Z \subseteq X$ we let $y(Z)$ denote $\sum_{i \in Z} y_i$.

A matching of a graph $G = (V, E, \psi)$ is a vector $x = (x_j : j \in E) \in \mathbb{R}^E$ satisfying

$$x_j = 0 \text{ or } 1 \text{ for all } j \in E ,$$

$$x(\delta(i)) \le 1 \text{ for all } i \in V .$$

Thus a matching can be thought of as being the incidence vector of a set of edges which meets each node at most once.

We define the matching polyhedron $P(G)$ to be the convex hull of the set of matchings of G. (See Stoer Witzgall [3] for general polyhedral theory). It has been shown (Edmonds [2]) that

$$P(G) = \{x = (x_j : j \in E) \in \mathbb{R}^E :$$

(1) $x_j \ge 0$ for all $j \in E$,

(2) $x(\delta(i)) \le 1$ for all $i \in V$,

(3) $x(\gamma(S)) \le \frac{|S| - 1}{2}$ for all $S \in Q\}$,

where $Q = \{S \subseteq V: |S| \text{ is odd}, |S| \ge 3\}$. However this set of inequalities (1), (2), (3) is generally far from minimal; in this paper we prescribe the unique minimal subset of these inequalities which serve to define $P(G)$.

If $(x^i \in \mathbb{R}^E : i \in I)$ is a family of vectors we say that they are <u>affinely independent</u> if whenever $\sum_{i \in I} \alpha_i x^i = 0$ and $\sum_{i \in I} \alpha_i = 0$ for some $(\alpha_i \in \mathbb{R} : i \in I)$ then we have $\alpha_i = 0$ for all $i \in I$. Notice that if a family of vectors is linearly independent then it is affinely independent. Let $X \subseteq \mathbb{R}^E$. Let I be an affinely independent subset of

X which is as large as possible. The <u>dimension</u> of X is defined to be $|I| - 1$.

It is well known that

(4) any subset of $\mathbb{R}^E$ has dimension at most $|E|$.

A polyhedron $P \subset \mathbb{R}^E$ is said to be <u>full dimensional</u> if P has dimension $|E|$.

(5) <u>Proposition</u>. <u>$P(G)$ is full dimensional</u>.

<u>Proof</u>. For each $j \varepsilon E$ let x^j be the matching defined by

$$x_k^j = \begin{cases} 0 & \text{if } k \neq j \\ 1 & \text{if } k = j. \end{cases}$$

Let x^0 denote the zero matching (i.e. $x_k^0 = 0$ for all $k \varepsilon E$).

Suppose (6) $\sum_{j \varepsilon E} \alpha_j x^j + \alpha_0 x_0 = 0$

and

(7) $\sum_{j \varepsilon E} \alpha_j + \alpha_0 = 0$.

From (6) we have $\sum_{j \varepsilon E} \alpha_j x^j = 0$ so since for all $j \varepsilon E$, x^j is the unique member of $\{x^k : k \varepsilon E\}$ whose j component is nonzero, we have $\alpha_j = 0$ for all $j \varepsilon E$ and so (7) implies $\alpha_0 = 0$ as well. Hence $P(G)$ contains $|E| + 1$ affinely independent vectors which together with (4) implies that $P(G)$ is full dimensional.

Since $P(G)$ is full dimensional there is a unique minimal subset of the inequalities (1), (2), (3) which will serve to define $P(G)$. (This uniqueness need not be the case for a polyhedron which is not of full dimension.) We call these irredundant inequalities the <u>facets</u> of $P(G)$. The purpose of this paper is to characterize the facets of $P(G)$.

A main tool which we use is the following theorem.

(8) <u>Theorem.</u> <u>Let P be a polyhedron of dimension n and let $ax \leq \alpha$ be one of the inequalities which define P. Then $ax \leq \alpha$ is a facet of P if and only if there are n affinely independent members x of P which satisfy $ax = \alpha$.</u>

(9) <u>Theorem.</u> <u>$x_j \geq 0$ is a facet of $P(G)$ for all $j \in E$.</u>

<u>Proof</u>. Let $j \in E$ and let P_j be the polyhedron defined by (2), (3) and

$$x_k \geq 0 \quad \text{for all } k \in E-\{j\}.$$

We define x^0 by

$$x_k^0 = \begin{cases} 0 & \text{if } k \in E-\{j\}, \\ -1 & \text{if } k = j \,. \end{cases}$$

Then $x^0 \in P_j - P(G)$, hence $x_j \geq 0$ is essential for defining $P(G)$ and so is a facet of $P(G)$.

Now we characterize which of the inequalities (2) are facets of $P(G)$. For any $i \in V$ we let $N(i)$ denote the set of nodes of G adjacent to i. If H is a connected component of G such that $|V(H)| = 2$, we call H a balanced edge. Thus a balanced edge consists of two nodes and one or more edges joining these nodes.

(10) Theorem $x(\delta(i)) \leq 1$ is a facet of $P(G)$ if and only if

(11) i is a node of a balanced edge

or

(12) $|N(i)| > 1$, and if $|N(i)| = 2$ then $\gamma(N(i)) = \phi$.

Proof. First we prove the necessity of (11) and (12). Suppose that $i \in V$ fails to satisfy (11) and (12). We will show that

(13) $$x(\delta(i)) \leq 1$$

is implied by other constraints (1), (3) and

(14) $$x(\delta(v)) \leq 1 \quad \text{for all} \quad v \in V-\{i\}$$

and so then (13) cannot be a facet of $P(G)$.

First suppose $N(i) = \{w\}$ for some $w \in V$. Then $\delta(i) \subseteq \delta(w)$ so (13) is implied by

(15) $$x(\delta(w)) \leq 1$$

together with (1). Since i violates (11), there is some $j \in \delta(w) - \delta(i)$ so (15) is a different inequality from (13). Thus (13) is not a facet of $P(G)$.

If $N(i) = \{w, u\}$ for distinct $w, u \in V$ then since $\delta(i) \subseteq \gamma\{i, w, u\})$ (13) is implied by the constraint

(16) $$x(\gamma(\{i, w, u\})) \leq 1$$

together with (1). Since i violates (12), there is some $j \in \gamma(\{w, u\})$. Thus $j \in \gamma(\{i, w, u\}) - \delta(i)$ and (16) is a different constraint from (13). Consequently (13) is not a facet of $P(G)$.

We now prove the sufficiency of condition (11) or (12). Suppose i is a node of a balanced edge. Let $j \in \delta(i)$. For each $h \in \delta(i)$ we define the matching x^h by

$$x_k^h = \begin{cases} 1 & \text{if } k = h, \\ 0 & \text{if } k \in E-\{h\}. \end{cases}$$

For each $h \in E - \delta(i)$ we define a matching x^h by

$$x_k^h = \begin{cases} 1 & \text{if } k = h \\ 1 & \text{if } k = j \\ 0 & \text{if } k \in E-\{h, j\}. \end{cases}$$

Clearly, the set $\{x^h : h \in E\}$ is linearly independent, and $x^h(\delta(i)) = 1$ for every $h \in E$ so by (8) we have that $x(\delta(i)) \leq 1$ is a facet of $P(G)$.

Suppose i satisfies (12). For every $h \in \delta(i)$ we define the matching x^h by

$$x_k^h = \begin{cases} 0 & \text{if } k \in E-\{h\}, \\ 1 & \text{if } k = h. \end{cases}$$

Now suppose $h \in E - \delta(i)$. If $|\psi(h) \cap N(i)| = 2$ then we must have $|N(i)| \geq 3$ else we would fail to satisfy (12). If $|\psi(h) \cap N(i)| \leq 1$ then there is some $v \in N(i) - \psi(h)$. Therefore for each $h \in E - \delta(i)$ there is some $j(h) \in \delta(i)$ such that $\psi(h) \cap \psi(j(h)) = \phi$.

For each $h \in E - \delta(i)$ we define a matching x^h of G by

$$x_k^h = \begin{cases} 1 & \text{if } k = h \\ 1 & \text{if } k = j(h) \\ 0 & \text{if } k \in E - \{h, j(h)\}. \end{cases}$$

These matchings are easily seen to be linearly independent and $x^h(\delta(i)) = 1$ for all $h \in E$, and so by (8), $x(\delta(i)) \leq 1$ is a facet of $P(G)$. This completes the proof of (10).

Now we turn to the problem of characterizing which of the constraints (3) are facets of $P(G)$. If H is a graph for which $|V(H)|$ is odd then obviously there can exist no perfect matching of H, i.e., a matching x for which $x(\delta(i)) = 1$ for all $i \in V(H)$. The closest we can possibly come is a matching x for which

$$x(\delta(i)) = 1 \quad \text{for all} \quad i \in V(H)-\{v\},$$

$$\text{and} \quad x(\delta(v)) = 0$$

for some $v \in V(H)$. We call such a matching a <u>near perfect matching</u> (<u>np matching</u>) of H which is <u>deficient</u> at v. If for every $i \in V(H)$ there is a np matching deficient at i then we say that H is <u>hypomatchable</u>. We call a hypomatchable graph H <u>degenerate</u> if $|V(H)| = 1$ (and hence $E(H) = \phi$); otherwise we call H <u>nondegenerate</u>.

A graph P is called an <u>odd polygon</u> if $|V(P)|$ is odd, P is connected and $|\delta(i)| = 2$ for each $i \in V(P)$. It is easily seen that odd polygons are hypomatchable. Moreover, every hypomatchable graph can be composed from odd polygons in a way which we now describe.

Let $G = (V, E, \psi)$ be a graph and let $S \subseteq V$. We define the graph obtained from G by shrinking S to be the graph $\bar{G} = (\bar{V}, \bar{E}, \bar{\psi})$ where

$$\bar{V} = V - S \cup \{S\},$$

$$\bar{E} = E - \gamma(S)$$

$$\bar{\psi}(j) = \begin{cases} \psi(j) & \text{if } j \in \bar{E} - \delta(S), \\ \psi(j) - S \cup \{S\} & \text{if } j \in \delta(S). \end{cases}$$

We denote $\bar{G}$ by $G \times S$ and call S a <u>pseudonode</u> of $\bar{G}$ (with respect to G).

Let $\mathcal{S}$ be a set of subsets of V. We say that $\mathcal{S}$ is <u>nested</u> if

$|S| \geq 3$ for every $S \in \mathcal{S}$, and for any $S, T \in \mathcal{S}$ such that $S \cap T \neq \phi$ we have $S \subset T$ or $T \subset S$.

For any $S \in \mathcal{S}$ let $\mathcal{S}[S] = \{T \in \mathcal{S} : T \subset S\}$. Finally, if $\{S_1, S_2, \ldots, S_k\}$ is the set of maximal members of $\mathcal{S}$, then we let $G \times \mathcal{S}$ denote $(\ldots((G \times S_1) \times S_2)\ldots \times S_k)$. It is easily seen that the order of the sets $S_1, S_2, \ldots, S_k$ has no effect on $G \times \mathcal{S}$.

If H is a subgraph of G such that $V(H) = V(G)$ then we say that H <u>spans</u> G.

(19) <u>Theorem. A graph $G = (V, E, \psi)$ is hypomatchable if and only if there is a (possibly empty) nested family $\mathcal{S}$ of subsets of V such that</u>

(20) <u>For every $S \in \mathcal{S}$, $G[S] \times \mathcal{S}[S]$ is spanned by an odd polygon P_S, and</u>

(21) <u>$|V(G \times \mathcal{S})| = 1$</u>. (i.e., $V \in \mathcal{S}$).

We call such a family $\mathcal{S}$ a <u>shrinking family</u> of G.

Proof. An algorithm has been developed which will find a matching x^0 of a graph $G = (V, E, \psi)$ which maximizes $x(E)$ over all matchings x of G (Edmonds [1]). This so called blossom algorithm proves the necessity of a hypomatchable graph having a shrinking family since if the blossom algorithm is applied to a graph G for which $|V(G)|$ is odd, it will either find a shrinking family or else find a node v of G at which no np matching of G can be deficient (an "inner node of a Hungarian tree").

We show the sufficiency of these conditions by proving a somewhat stronger result:

(22) If $\mathcal{S}$ is a shrinking family of $G = (V, E, \psi)$ then for every $v \in V$ there is a np matching x^v of G deficient at v such that

(23) for every $S \in \mathcal{S}$, $x^v|\gamma(S)$ is a np matching of $G[S]$.

We prove by induction on $|\mathcal{S}|$. If $|\mathcal{S}| = 0$ then G is degenerate and (22) is trivial. Assume it is true for shrinking families of fewer than n sets for $n > 0$ and assume $|\mathcal{S}| = n$. By (21) $V \in \mathcal{S}$. Let v be any node of V. Every maximal $S \in \mathcal{S}[V]$ is a pseudonode of the odd polygon P_V, let $p = v$ if $v \in V(P_V)$, let $p = S$ if v is contained in pseudonode S of P_V. Let $\tilde{x}$ be a np matching of P_V deficient at p. Now for every pseudonode $T \in V(P_V)$ there will be at most one node of T incident with some $j \in E(P_V)$ for which $\tilde{x}_j = 1$. If such a node, say $w(T)$, exists, let $\bar{x}^T$ be a np matching of $G[T]$ deficient at $w(T)$ such that

(24) $\bar{x}^T|\gamma(S)$ is a np matching of $G[S]$

for every $S \in \mathcal{S}[T]$,

which exists by our induction hypothesis. If no such $w(T)$ exists then $v \in T$ and we let $\bar{x}^T$ be a np matching of $G[T]$ deficient at v which satisfies (24). Now we define x^v by

$$x_j^v = \begin{cases} \bar{x}_j & \text{for } j \in E(P_v) \\ 0 & \text{for } j \in E(G \times \mathcal{S}[V]) - E(P_v) \\ \bar{x}_j^T & \text{for } j \in \gamma(T), \text{ for all } T \in \mathcal{S}[V]. \end{cases}$$

It is easily seen that x^v is a np matching of G, deficient at v, which satisfies (23), and so (22) is proved. This also completes the proof of (19).

The following lemma is useful when proving the linear independence of a set of matchings.

(25) Lemma. Let X be a set of np matchings of $G = (V, E, \psi)$ and let $x^0 \in X$. If there are $J(x^0) \subseteq E$ and $d(x^0) \in \mathbb{R}$ such that $x^0(J(x^0)) < d(x^0)$ but $x(J(x^0)) = d(x^0)$ for all $x \in X - \{x^0\}$ then x^0 is linearly independent of $X - \{x^0\}$.

Proof. Suppose that there are $\alpha_x \in \mathbb{R}$ for $x \in X' = X - \{x^0\}$ such that

(26) $$x^0 = \sum_{x \in X'} \alpha_x x .$$

Since every member of X is a np matching of G, the value of $x(E)$ is the same for every $x \in X$. By (26), we have

$$x^0(E) = \sum_{x \in X'} \alpha_x \, x(E), \text{ and hence}$$

(27) $$\sum_{x \in X'} \alpha_x = 1 .$$

Therefore $\sum_{x \in X'} \alpha_x x(J(x^0)) = \sum_{x \in X'} \alpha_x d(x^0) = d(x^0)$ by (27). Hence (26) implies that $x^0(J(x^0)) = d(x^0)$, a contradiction which proves the lemma.

A <u>cutnode</u> of $G = (V, E, \psi)$ is a node v such that $G[V - \{v\}]$ has more connected components than G. A <u>block</u> B of G is a maximal connected subgraph of G such that B has no cutnode. We let $b(G)$ denote the number of blocks of G. It is easily seen that

(28) if x is a np matching of hypomatchable G then $x|E(B)$ must be a np matching of each block B of G.

We now prove a main theorem used for characterizing which of the constraints (3) are facets of $P(G)$.

(29) <u>Theorem</u>. <u>If $G = (V, E, \psi)$ is hypomatchable then G has $|E| - (b(G) - 1)$ linearly independent np matchings</u>.

<u>Proof</u>. Let $\mathcal{S}$ be a shrinking family of G; we prove by induction on $|\mathcal{S}|$. If $|\mathcal{S}| = 0$ then G is degenerate, $|E| = 0$, $b(G) = 1$ and the result is trivial. Suppose (29) is true whenever the shrinking family has fewer than m members for $m > 0$ and that $|\mathcal{S}| = m$.

Let P be an odd polygon which spans $G \times \mathcal{S}[V]$ and exists by (20). We partition $V(P)$ into $V_1 \cup V_2$ where $V_1 = V(P) \cap V$ and $V_2 = V(P) \cap \mathcal{S}$ (that is, V_1 is the set of real nodes and V_2 is the set of pseudonodes of P).

Let $C = E(G \times \mathcal{S}[V]) - E(P)$ and let $G' = (V, E', \psi')$ be the graph obtained from G by deleting all those edges of C. Then $\mathcal{S}$ is a shrinking family of G' and so by (22) for each $v \in V_1$ there is a np matching $\bar{x}^v$ of G' deficient at v which satisfies (23) for G'. For each $v \in V_1$ we define a np matching x^v of G deficient at v by

$$(30) \qquad x_j^v = \begin{cases} \bar{x}_j^v & \text{for } j \in E' \\ 0 & \text{for } j \in C. \end{cases}$$

Let $X_1 = \{x^v : v \in V_1\}$. Since by (23) each $x \in X$ is a np matching of $G[S]$ for each $S \in V_2$, it follows from (28) that

(31) $x|E(B)$ is a np matching of B for every block B of $G[S]$, for every $S \in V_2$, for every $x \in X_1$.

For each $S \in V_2$ there are by induction $n(S) = |\gamma(S)| - (b(G[S]) - 1)$ linearly independent np matchings $\{\bar{x}^{S,1}, \bar{x}^{S,2}, \ldots, \bar{x}^{S,n(S)}\}$ of $G[S]$ since $\mathcal{S}[S] \cup \{S\}$ is a shrinking family of $G[S]$ and $|\mathcal{S}[S] \cup \{S\}| \leq |\mathcal{S} - \{V\}| < m$. By (28)

(32) $\bar{x}^{S,i}|E(B)$ is a np matching of B for every block B of $G[S]$, for every $i \in \{1, 2, \ldots, n(S)\}$. We extend each to a np matching of G as follows.

Let $\tilde{x}^S$ be the np matching of P deficient at S. For each $T \in V_2 - \{S\}$ let $j_T \in \delta(T)$ be the edge such that $\tilde{x}^S_{j_T} = 1$, let $\{v_T\} = \psi(j_T) \cap T$ and let $\bar{x}^{T,S}$ be a np matching of $G[T]$ deficient at v_T. Then by (28)

(33) $\bar{x}^{T,S}|\gamma(B)$ is a np matching of B for every block B of $G[T]$.

Now we define $x^{S,i}$ for all $i \in \{1, 2,\ldots,n(S)\}$ by

$$(34)\quad x^{S,i}_j = \begin{cases} \bar{x}^{S,i}_j & \text{for } j \in \gamma(S), \\ \tilde{x}^S_j & \text{for } j \in E(P), \\ 0 & \text{for } j \in C \\ \bar{x}^{T,S}_j & \text{for all } j \in \gamma(T), \text{ for all } T \in V_2 - \{S\}. \end{cases}$$

Let $X_2 = \{x^{S,i}: i \in \{1, 2,\ldots,n(S)\}, S \in V_2\}$. By (32) and (33) we have

(35) $x|E(B)$ is a np matching of B for every block B of $G[T]$, for every $T \in V_2$, for every $x \in X_2$.

Now we show

(36) $X_1 \cup X_2$ is linearly independent.

Suppose $\alpha_v \in \mathbb{R} : v \in V_1$ and $\alpha_{S,i} \in \mathbb{R}: i \in \{1, 2,\ldots,n(S)\}$, $S \in V_2$, are such that

$$(37)\quad \sum_{v \in V_1} \alpha_v x^v + \sum_{S \in V_2} \sum_{i=1}^{n(S)} \alpha_{S,i} x^{S,i} = 0.$$

If we let $\tilde{x}^v = x^v|E(P)$ for all $v \in V_1$ we have

$$\sum_{v \in V_1} \alpha_v \tilde{x}^v + \sum_{S \in V_2} \bar{\alpha}_S \tilde{x}^S = 0$$

where

$$\bar{\alpha}_S = \sum_{i=1}^{n(S)} \alpha_{S,i} \quad \text{for every} \quad S \in V_2 .$$

For each $v \in V(P)$, $\tilde{x}^v$ is a np matching of P deficient at v so if we let $J(\tilde{x}^v) = \delta(v) \cap E(P)$ and $d(\tilde{x}^v) = 1$ for all $v \in V(P)$ we have by (25) that $\{\tilde{x}^v: v \in V(P)\}$ is linearly independent so

(38) $\quad \alpha_v = 0$ for all $v \in V_1$,

(39) $\quad \bar{\alpha}_S = 0$ for all $S \in V_2$.

Now let $S \in V_2$, let $V_2' = V_2 - \{S\}$. By (34), (37) and (38) we have

$$\sum_{i=1}^{n(S)} \alpha_{S,i} \bar{x}^{S,i} + \sum_{T \in V_2'} \bar{\alpha}_T \bar{x}^{S,T} = 0$$

so by (39)

$$\sum_{i=1}^{n(S)} \alpha_{S,i} \bar{x}^{S,i} = 0 \quad .$$

But the matchings $\{\bar{x}^{S,i}: i \in \{1, 2, \ldots, n(S)\}\}$ are by hypothesis linearly independent so

(40) $\quad \alpha_{S,i} = 0$ for all $i \in \{1, 2, \ldots, n(S)\}$.

Thus (38) and (40) together prove (36).

Let $k \in C$. We define a np matching x^k as follows. Let v and w be the nodes of P met by k, let $\tilde{x}^v$ be the np matching of P deficient at v. There must be some $\ell \in E(P) \cap \delta(w)$ such that $\tilde{x}^v_\ell = 1$, we define a np matching $\hat{x}^k$ of $G \times \mathcal{S}[V]$ by

$$\hat{x}^k_j = \begin{cases} \tilde{x}^v_j & \text{for } j \in E(P) - \{\ell\} \\ 0 & \text{if } j \in (C - \{k\}) \cup \{\ell\} \\ 1 & \text{if } j = k . \end{cases}$$

Now let $T \in V_2$. If $\hat{x}^k(\delta(T)) = 0$ we let $\bar{x}^T$ be any np matching of $G[T]$. If there is some $\ell \in \delta(T)$ such that $\hat{x}^k_\ell = 1$ then let $\{v\} = \psi(\ell) \cap T$ and let $\bar{x}^T$ be a np matching of $G[T]$ which is deficient at v.

Now we define x^k by

$$x^k_j = \begin{cases} \hat{x}^k_j & \text{for } j \in E(G \times \mathcal{S}[V]) \\ \bar{x}^T_j & \text{for } j \in \gamma(T) \text{ for } T \in V_2 . \end{cases}$$

Let $X_3 = \{x^k\colon k \in C\}$. Every member of x of X_3 is a np matching of G and $x|\gamma(T)$ is a np matching of $G[T]$ for all $T \in V_2$, so by (28)

(41) $x|E(B)$ is a np matching of every block B of $G[S]$ for every $S \in V_2$ for every $x \in X_3$.

Moreover, since by (30) and (34) for each $k \in C$, x^k is the unique member x of $X_1 \cup X_2 \cup X_3$ such that $x_k \neq 0$, we have using (36)

(42) $X_1 \cup X_2 \cup X_3$ is linearly independent

Now let B be a block of $G[S]$ such that B is not a block of G for some $S \in V_2$. Let $\{h, k\} = E(P) \cap \delta(S)$, let $\{v\} = \psi(h) \cap S$, let $\{w\} = \psi(k) \cap S$.

A path in G joining node v_0 to node v_n is a sequence $(v_0, j_1, v_1, j_2, v_2, \ldots, j_n, v_n)$ where $v_i \in V$ for $i \in \{0, 1, \ldots, n\}$ and $j_i \in E$ for $i \in \{1, 2, \ldots, n\}$ such that $\psi(j_i) = \{v_i, v_{i-1}\}$ for $i \in \{1, 2, \ldots, n\}$ and for any $i, \ell \in \{0, 1, \ldots, n\}$ we have $v_i \neq v_\ell$.

We distinguish two cases.

<u>Case 1</u>. <u>Every path in $G[S]$ joining v to w contains an edge of B</u>. Since B is a block of $G[S]$ there is a unique node $p \in V(B)$ which is the first node of B in each such path. Let $\bar{H}$ be the connected component of $G[S - \{p\}]$ such that $V(B) - \{p\} \subseteq V(\bar{H})$. Let $H = G[V(\bar{H}) \cup \{p\}]$; let $K = G[S - V(\bar{H})]$. (K may just consist of the single node p). Then $S = V(H) \cup V(K)$ and $V(H) \cap V(K) = \{p\}$. It can be seen that any block of $G[S]$ is a subgraph of K or of H, and hence, using (28), that K and H are hypomatchable. Clearly $v \in V(K)$. By the hypothesis for Case 1, we have $p \neq w$ and $w \in V(H)$. Let x^H be a np matching of H deficient at w. Since $p \neq w$, there must be some $\ell \in E(B) \cap \delta(p)$ such that $x^H_\ell = 1$. Let x^K be a np matching of K deficient at v. Let $\{t\} = \psi(k) - S$; let u be the node of $V(P) - \{S\}$ met by k. Let $\tilde{x}$ be the np matching of P deficient at u. Then $\tilde{x}_h = 1$ and

(43) for each $T \in V_2 - \{u\}$ let $j(T)$ be the unique edge j of $\delta(T) \cap E(P)$ such that $\tilde{x}_j = 1$ and let x^T be a np matching of $G[T]$ deficient at $v(T)$ where $\{v(T)\} = \psi(j(T)) \cap T$.

If $u \in V_2$ then

(44) let x^u be a np matching of $G[u]$ deficient at t.

Now we define a np matching x^B of G by

$$(45)\quad x_j^B = \begin{cases} x_j^H & \text{for } j \in E(H) - \{\ell\}, \\ 0 & \text{for } j \in \{\ell\} \cup C, \\ x_j^K & \text{for } j \in E(K), \\ \tilde{x}_j & \text{for } j \in E(P) - \{k\}, \\ 1 & \text{for } j = k, \\ x_j^T & \text{for } j \in \gamma(T),\ \text{for } T \in V_2 - \{S\}. \end{cases}$$

It can now be seen that

(46) $x^B|E(D)$ is a np matching of D for every block D of every $G[T]$ for $T \in V_2$ unless $D = B$

and

$$(47)\qquad x^B(E(B)) = \frac{|V(B)| - 3}{2}\ .$$

<u>Case 2. No path in $G[S]$ joining v to w contains an edge of B.</u> (Since B is a block of $G[S]$, cases 1

and 2 exhaust all possibilities). There must exist $p \in V(B)$ such that every path in $G[S]$ from a node of B to v or w contains p. (p will be a cutnode of $G[S]$ unless $v = w$ and $v \in V(B)$ in which case $p = v$.) Since B is not a block in G, there must be some $e \in C \cap \delta(S)$ such that if $\{q\} = \psi(e) \cap S$, there is a path from q to a node of B which does not contain p. Let $\bar{H}$ be the connected component of $G[S - \{p\}]$ which contains q, let $H = G[V(\bar{H}) \cup \{p\}]$, let $K = G[S - V(\bar{H})]$. (If $v = w = p$ then K may simply contain the single node p). Let u be the node of $V(P) - \{S\}$ met by e and let $\tilde{x}$ be the np matching of P deficient at u. Let x^H be a np matching of H deficient at q. There must be some $\ell \in E(B) \cap \delta(p)$ such that $x^H_\ell = 1$. Let x^K be a np matching of K deficient at v or w, according to whether $\tilde{x}_h = 1$ or $\tilde{x}_k = 1$. For each $T \in V_2 - \{u\}$ define x^T as in (43). If $u \in V_2$, let $\{t\} = \psi(e) - S$ and define x^u as in (44). Now we define x^B as follows.

$$x^B_j = \begin{cases} x^H_j & \text{for } j \in E(H) - \{\ell\}, \\ 0 & \text{for } j \in C \cup \{\ell\} - \{e\}, \\ x^K_j & \text{for } j \in E(K), \\ \tilde{x}_j & \text{for } j \in E(P), \\ 1 & \text{for } j = e, \\ x^T_j & \text{for } j \in \gamma(T) \text{ for } T \in V_2 - \{S\}. \end{cases}$$

Now it can be seen that x^B is a np matching of G

satisfying (46) and (47).

Now let $\mathcal{D}$ = {blocks B of $G[S]$ for $S \in V_2$: B is not a block of G} and let $X_4 = \{x^B : B \in \mathcal{D}\}$. We have

$$x^B \text{ satisfies (46) and (47) for every } x^B \in \mathcal{D}.$$

Therefore by lemma (25), (32), (35), (41) and (42) we have that

$$X_1 \cup X_2 \cup X_3 \cup X_4 \text{ is linearly independent.}$$

Now we evaluate $|X_1 \cup X_2 \cup X_3 \cup X_4|$. Let $\mathcal{B}$ = {blocks of $G[S]$ for all $S \in V_2$}. Then we have

$$(48) \quad |X_1 \cup X_2 \cup X_3 \cup X_4|$$

$$= |V_1| + \sum_{S \in V_2} (|\gamma(S)| - b(G[S]) + 1) + |C| + |\mathcal{D}|$$

$$= |V_1| + |V_2| + \sum_{S \in V_2} |\gamma(S)| + |C| - (|\mathcal{B}| - |\mathcal{D}|).$$

Since $V_1 \cup V_2 = V(P)$ and $|V(P)| = |E(P)|$ and since E partitions into $E(P) \cup C \cup \bigcup_{S \in V_2} \gamma(S)$,

$$(49) \quad |E| = |V_1| + |V_2| + \sum_{S \in V_2} |\gamma(S)| + |C|.$$

Since the blocks of G consist of the members of $\mathcal{B} - \mathcal{D}$ together with the block containing $E(P)$, we have

$$(50) \qquad |\mathcal{B}| - |\mathcal{D}| = b(G) - 1.$$

Thus (48), (49), and (50) combine to give

$$|X_1 \cup X_2 \cup X_3 \cup X_4| = |E| - (b(G) - 1)$$

and (29) is proved.

We now characterize which of the inequalities (3) are facets of $P(G)$.

(51) Theorem. $x(\gamma(S)) \le \frac{1}{2}(|S| - 1)$ is a facet of $P(G)$ for $S \in Q$ if and only if

(52) $G[S]$ is hypomatchable and contains no cutnode

or

(53) $|S| = 3$ and $\gamma(S) = \delta(i)$ for some $i \in S$ and i satisfies (11) or (12).

Proof. Suppose $G[S]$ is hypomatchable and contains no cutnode. A matching x of G satisfies

$$(54)\quad x(\gamma(S)) = \frac{|S|-1}{2}$$

if and only if $x|\gamma(S)$ is a np matching of $G[S]$. By (29), $G[S]$ has a set $\hat{X}_1$ of $|\gamma(S)|$ linearly independent np matchings. We extend each $\hat{x} \in \hat{X}_1$ to a matching x of G by letting

$$(55)\quad x_j = \begin{cases} 0 & \text{if } j \in E - \gamma(S) \\ \hat{x}_j & \text{if } j \in \gamma(S). \end{cases}$$

Let X_1 be the set of matchings thereby obtained; each $x \in X_1$ satisfies (54).

Let $k \in \delta(S)$, let $\{v\} = \psi(k) \cap S$. Let $\tilde{x}$ be a np matching of $G[S]$ deficient at v and let x^k be defined by

$$(56)\quad x_j^k = \begin{cases} \tilde{x}_j & \text{for } j \in \gamma(S) \\ 1 & \text{for } j = k \\ 0 & \text{for } j \in E - (\gamma(S) \cup \{k\}). \end{cases}$$

For any $k \in \gamma(V - S)$ let $\tilde{x}$ be any np matching of $G[S]$ and let x^k be defined as in (56). Let $X_2 = \{x^k : k \in E - \gamma(S)\}$. Each $x^k \in X_2$ satisfies (54) and since by (55) and (56) x^k is the unique member x of $X_1 \cup X_2$ for which $x_k \neq 0$,

$X_1 \cup X_2$ is linearly independent.

Since $|X_1 \cup X_2| = |E|$, we have by (8) that

$$(57)\qquad x(\gamma(S)) \leq \tfrac{1}{2}(|S| - 1)$$

is a facet of $P(G)$.

If $|S| = 3$ and $\gamma(S) = \delta(i)$ for some $i \in S$ then (57) is the same as the constraint $x(\delta(i)) \leq 1$ which by (10) is a facet of $P(G)$ if and only if (11) or (12) is satisfied.

Suppose $S \subseteq V$ and $G[S]$ is not hypomatchable, and if $|S| = 3$ then $\gamma(S) \neq \delta(i)$ ror any $i \in S$. Then the fact that $x(\gamma(S)) \leq \frac{1}{2}(|S| - 1)$ is not a facet of $P(G)$ is a direct consequence of (19) and the blossom algorithm (Edmonds [2]) for finding a maximum weighted matching. It is implicit in this algorithm that the only constraints (3) which are required are those for which $G[S]$ has a shrinking family. However this can also be proved from the fact that $P(G) = \{x \in \mathbb{R}^E)$ satisfying (1), (2), (3)} as we show in (64).

Finally, suppose $G[S]$ is hypomatchable but $G[S]$ contains a cutnode. Let X, Y be nonempty subsets of S such that $S = X \cup Y$, $|X \cap Y| = 1$, $\gamma(X) \cup \gamma(Y) = \gamma(S)$. It follows from (28) that $G[X]$ and $G[Y]$ must be hypomatchable, hence $|X|$ and $|Y|$ are odd and there are constraints (3)

$$(58) \qquad x(\gamma(X)) \le \frac{|X|-1}{2}$$

$$(59) \qquad x(\gamma(Y)) \le \frac{|Y|-1}{2}$$

Adding (58) and (59) gives

$$x(\gamma(S)) \le \frac{|X| + |Y| - 2}{2}$$

$$= \frac{|S| - 1}{2}$$

so $x(\gamma(S)) \le \frac{1}{2}(|S|-1)$ is not a facet of $P(G)$. The proof of (51) is now complete.

Whenever we know a set of linear inequalities whose solution set is a given polyhedron P, linear programming duality enables us immediately to obtain a min-max theorem regarding the members of P. Thus the theorem that $P(G) = \{x \in \mathbb{R}^E$: x satisfies (1), (2), (3)} is equivalent to the following.

(60) Theorem. Let $c = (c_j : j \in E) \in \mathbb{R}^E$. Then the maximum value of $\sum_{j \in E} c_j x_j$ for all matchings x of $G = (V, E, \psi)$ is equal to the minimum value of

$$\sum_{i \in V} y_i + \sum_{S \in Q} \tfrac{1}{2}(|S| - 1) y_S \tag{61}$$

for $y \in \mathbb{R}^{V \cup Q}$ which satisfy

$$\begin{aligned} y_i &\geq 0 \quad \text{for all } i \in V, \\ y_S &\geq 0 \quad \text{for all } S \in Q \end{aligned} \tag{62}$$

$$y(\psi(j)) + y(Q(j)) \geq c_j \quad \text{for all } j \in E \tag{63}$$

where for each $j \in E$, $Q(j) = \{S \in Q: j \in \gamma(S)\}$.

It is a direct consequence of (60) that for any $S \in Q$, $x(\gamma(S)) \leq \frac{1}{2}(|S| - 1)$ is not a facet of $P(G)$ if S violates (52) and (53).

(64) Proof. If $G[S]$ is not hypomatchable then there is $v \in S$ such that $G[S - \{v\}]$ has no perfect matching. Let $S' = S - \{v\}$. Then the maximum value of $x(\gamma(S'))$ for matchings x of $G[S']$ is at most $\frac{1}{2}(|S| - 3)$. Therefore by (60) there is $y^0 \in \mathbb{R}^{S' \cup Q'}$ where $Q' = \{T \subseteq S': |T| \text{ is odd}, |T| \geq 3\}$ such that

$$y^0 \geq 0, \tag{65}$$

$$y^0(\psi(j)) + y^0(Q'(j)) \geq 1 \quad \text{for all } j \in \gamma(S') \tag{66}$$

$$\sum_{i \in S'} y_i^0 + \sum_{T \in Q'} \tfrac{1}{2}(|T| - 1) y_T^0 \leq \tfrac{1}{2}(|S| - 3) \tag{67}$$

If we multiply each inequality $x(\delta(i)) \le 1$ by y_i^0 for $i \in S'$ and multiply each inequality $x(\gamma(T)) \le \frac{1}{2}(|T|-1)$ by y_T^0 for $T \in Q'$ and add these inequalities together we obtain an inequality

$$(68) \qquad \sum_{j \in J} a_j x_j \le b \ .$$

By (66) we have $a_j \ge 1$ for all $j \in \gamma(S')$ and by (67) $b \le \frac{1}{2}(|S| - 3$. Hence (68) implies

$$x(\gamma(S')) \le \frac{1}{2}(|S| - 3).$$

If we add to this the inequality $x(\delta(v)) \le 1$ we obtain

$$(69) \qquad x(\gamma(S)) \le \frac{1}{2}(|S| - 1).$$

None of the inequalities used to obtain (69) could be the same as (69) unless $|S| = 3$ and $\gamma(S) = \delta(i)$ for some $i \in S$ which we do not allow. This completes the proof.

It is a consequence of the blossom algorithm that

(70) if c_j is integer valued for all $j \in E$ then theorem (60) holds if we add the restrictions

$$(71) \qquad y_S \equiv 0 (\text{mod } 1) \text{ for } S \in Q,$$

$$(72) \qquad y_i \equiv 0 (\text{mod } \tfrac{1}{2}) \text{ for } i \in V.$$

This is turn can be seen to imply

(73) if $c_j \in \{0, 1\}$ for all $j \in E$ then (60) holds if we require (71) and

$$(74) \qquad y_i \equiv 0 (\text{mod } 1) \text{ for } i \in V.$$

As a result of (9), (10) and (51) we can obtain a "best possible" strengthening of (60) by replacing V and Q in (61)-(63) with

$$V^* = \{i \in V: i \text{ satisfies } (12)\} \cup W$$

where $W \subseteq V$ contains exactly one node from each balanced edge and

$$Q^* = \{S \in Q: S \text{ satisfies } (52)\}.$$

This is best possible for if we replaced either V^* or Q^* by a smaller set, then the min-max relation would no longer be true for all $c \in \mathbb{R}^E$. It can be seen that (70) and (73) also hold for this stronger theorem.

Matching theory for nonbipartite graphs developed from a theorem of Tutte [4] giving necessary and sufficient conditions for a graph to have a perfect matching. Both this theorem and a min-max theorem of Berge [5] can be easily derived from (60) together with (73). We now discuss a characterization of hypomatchable graphs which is closely related to this theorem of Tutte.

If $G = (V, E, \psi)$ is a graph, we let $c(G)$ denote the number of connected components H of G such that $|V(H)|$ is odd. We call these <u>odd components</u>, other components are called <u>even components</u>.

(75) <u>Theorem</u> <u>(Tutte [4])</u>. <u>$G = (V, E, \psi)$ has a perfect matching if and only if for every $X \subseteq V$,</u>

(76) $$|X| \geq c(G[V - X]) \; .$$

If G is hypomatchable and hence $|V|$ is odd then $X = \phi$ trivially violates (76). We now show that G is hypomatchable if and only if $X = \phi$ is the only subset of V violating (76).

(77) <u>Theorem</u>. <u>$G = (V, E, \psi)$ is hypomatchable if and only if $|V|$ is odd and for every nonempty $X \subseteq V$,</u>

$$\underline{|X| \geq c(G[V - X])}.$$

<u>Proof</u>. Suppose G is hypomatchable, we have already noted that $|V|$ is odd. Let $X \subseteq V$ be nonempty, let x be a np matching of G deficient at $v \in X$. If H is any odd component of $G[V - X]$ then $x(\delta(V(H))) \geq 1$ since $x(\delta(i)) = 1$ for all $i \in V(H)$. Hence

(78) $$x(\delta(V - X)) \geq c(G[V - X])$$

Since $x(\delta(i)) \leq 1$ for all $i \in X$ we have

(79) $$x(\delta(X)) \leq |X|.$$

Since $\delta(X) = \delta(V - X)$, (78) and (79) combine to prove (76).

Now suppose G is not hypomatchable, but $|V|$ is odd. There is some $v \in V(G)$ at which no np matching is deficient, so $G' = G[V - \{v\}]$ has no perfect matching and so by (75) there is $X \subseteq V - \{v\}$ such that

$$|X| < c(G'[V - \{v\} - X]).$$

Since $|X|$ and $c(G'[V - \{v\} - X])$ must be of the same parity ($|V - \{v\}|$ being even) we have

(80) $|X| \leq c(G'[V - X - \{v\}]) - 2.$

Obviously $c(G[V - (X \cup \{v\})]) = c(G'[V - \{v\} - X])$ so by (80)

$$|X \cup \{v\}| < c(G[V - (X \cup \{v\})])$$

$X \cup \{v\} \neq \phi$ and the theorem is proved.

Of course theorems (51) and (77) can be combined to give a second characterization.

(81) Theorem. For any $S \in Q$, $x(\gamma(S)) \leq \frac{|S| - 1}{2}$ is a facet of $P(G)$ if and only if

$|S|$ is odd, $G[S]$ contains no cutnode and for every nonempty $X \subseteq S$, $|X| \geq c(G[S - X])$,

or

$|S| = 3$, $\gamma(S) = \delta(i)$ for some $i \in S$ and i satisfies (11) or (12).

Another problem to which we can apply these results is enumerating the number of linearly independent np matchings of a graph $G = (V, E, \psi)$ for which $|V|$ is odd. By elementary linear algebra, G can have at most $|E|$ such matchings. We can characterize those graphs which attain the maximum as follows.

(82) Theorem. Let $G = (V, E, \psi)$ be a graph for which $b(V)$ is odd. Then G has $|E|$ linearly independent np matchings if and only if G is hypomatchable and has no cutnode or $|V| = 3$.

Proof. By (8) G has $|E|$ linearly independent np matchings if and only if $x(E) \le \frac{|V|-1}{2}$ is a facet of $P(G)$. By (51) this holds if and only if G is hypomatchable and has no cutnode or if $|V| = 3$, $E = \delta(i)$ for some $i \in V$ and i satisfies (11) or (12). But this last condition can be seen to always hold when $|V| = 3$, completing the proof.

If we define the matching rank of a graph $G = (V, E, \psi)$ for which $|V|$ is odd to be the cardinality of the largest linearly independent set of np matchings then the following strengthening of (29) can be proved.

(83) Theorem. The matching rank of a hypomatchable graph $G = (V, E, \psi)$ is equal to $|E| + 1 - b(G)$.

Finally we note that all results presented here have been generalized to the case of b-matchings, matching problems where constraints of the form $x(\delta(i)) \le 1$ have been replaced by constraints of the form $x(\delta(i)) \le b_i$ for a positive integer b_i. These results will be reported elsewhere.

The results presented here together with the generalization to b-matchings are a part of the Ph.D. thesis of W. Pulleyblank prepared under the supervision of Professor Jack Edmonds at the University of Waterloo.

REFERENCES

1. J. Edmonds, "Paths, Trees and Flowers", Canadian J. Math. 17, 449-467 (1965).

2. J. Edmonds, "Maximum Matching and a Polyhedron with 0, 1-Vertices", J. Res. Nat. Bur. of Standards 69B, (Math and Math Phys.) No. 1, 125-130 (1965).

3. J. Stoer, C. Witzgall, Convexity and Optimization in Finite Dimensions I, Springer-Verlag, Berlin, Heidelberg (1970).

4. W.T. Tutte, "The Factorization of Linear Graphs", J. London Math. Soc. 22, 107-111 (1947).

5. C. Berge, "Sur le couplage maximum d'un graphe" C.R. Acad. Sci. Paris 247, 258-259 (1958).

CHROMIALS

W.T. Tutte, University of Waterloo

1. Introduction.

In their paper "Chromatic Polynomials", of 1946, ([2]), G. D. Birkhoff and D. C. Lewis advocated a quantitative approach to the Four Colour Problem. This would be concerned with the properties of a function $P(G, 4)$, defined as the number of 4-colourings of a given graph G. In the hope of greater generality they extended their discussion to $P(G, \lambda)$, the number of vertex-colourings of G in λ colours. This proved to be a polynomial in λ for each graph G. It could therefore be extended from integral to complex values of λ. Birkhoff and Lewis expressed the hope that eventually it would be possible to study the Four Colour Problem by applying methods of real or complex analysis to the "chromatic polynomials" $P(G, \lambda)$.

Let us start by defining $P(G, \lambda)$ precisely. For any positive integer λ we define a λ-colouring of a graph G as a mapping f of the vertex-set of G into the set I_λ of integers from 1 to λ, provided that f maps the ends of any edge of G onto two distinct members of I_λ. The members of I_λ are the λ colours. It is not necessary that every colour shall be assigned by f to a vertex of G. For example a 3-colouring can be regarded as a special kind of 4-colouring.

When G has a loop the stated condition cannot be satisfied for any λ, and such a graph has no λ-colouring. Only an edgeless graph can have a 1-colouring, and only graphs without odd circuits have 2-colourings. A trivial extension to the case $\lambda = 0$ can be made. Allowing that there is exactly one mapping of a null set onto a null set we can say that the null graph has exaclly one 0-colouring, but no other graph has a 0-colouring.

$P(G, \lambda)$ is by definition the number of λ-colourings of G. It is readily calculated for some graphs of exceptionally simple form, for example the edgeless graphs and the complete graphs K_n. The results, quite obvious, are as follows.

1.1. <u>If G is an edgeless graph of n vertices, then</u>

$$P(G, \lambda) = \lambda^n.$$

1.2. <u>If G is the complete graph K_n of n vertices, then</u>

$$P(G, \lambda) = \lambda(\lambda - 1)(\lambda - 2) \dots (\lambda - n + 1).$$

It is usual to interpret an empty product as unity. Hence the above formulae are consistent with our observation that $P(G, 0) = 1$ when G is a null graph.

There are two basic recursion formulae in the theory of $P(G, \lambda)$.

The first of these deals with the case in which G is the union of two subgraphs H and K such that the intersection H K is a complete graph K_n. It asserts that

$$1.3. \qquad P(G, \lambda) = \frac{P(H, \lambda)P(K, \lambda)}{P(H \cap K, \lambda)}.$$

The case n = 0 tells us that if G is the union of two disjoint subgraphs H and K, then

$$P(G, \lambda) = P(H, \lambda)P(K, \lambda).$$

In the case n = 1 the common part of H and K consists of a single vertex. We then have

$$P(G, \lambda) = \lambda^{-1}P(H, \lambda)P(K, \lambda).$$

To prove 1.3 in the general case we observe that a general λ-colouring of G can be obtained by first colouring H and then colouring K so as to agree with the colouring of H on all the common vertices. The colourings of K can be arranged in $P(H \cap K, \lambda)$ classes, according to the colourings they induce in $H \cap K$. Since $H \cap K$ is a complete graph each of these classes has the same number of members, for any class can be transformed into any other by a permutation of colours. Hence the number of λ-colourings of K agreeing with a chosen λ-colouring of H is $P(K, \lambda)/P(H \cap K, \lambda)$. Proposition 1.3 follows.

The second recursion formula concerns a link A of a graph G. By deleting A we transform G into a graph G'_A. If, besides deleting A, we identify its two ends to form a single new vertex z we obtain from G a graph G''_A. The formula is

1.4. $$P(G, \lambda) = P(G'_A, \lambda) - P(G''_A, \lambda).$$

<u>Proof</u>.

The λ-colourings of G'_A in which the two vertices x and y, incident with A in G, have different colours are the λ-colourings of G. Those in which x and y have the same colour are in 1 - 1 correspondence with the λ-colourings of G''_A.

We can base the theory of $P(G, \lambda)$ on the four following Axioms.

I. $P(G, \lambda) = 0$ <u>if G has a loop</u>.

II. $P(G, \lambda) = \lambda$ <u>if G is a vertex-graph</u>.

III. <u>If</u> $G = H \cup K$, <u>where H and K are disjoint, then</u>

$$P(G, \lambda) = P(H, \lambda)P(K, \lambda).$$

IV. <u>If A is a link of G then</u>

$$P(G, \lambda) = P(G'_A, \lambda) - P(G''_A, \lambda).$$

The axioms are contained in Propositions 1.1 to 1.4. A "vertex-graph" is a graph consisting of a single vertex and no edge.

For any given graph G these axioms either determine P(G, λ) at once or they express P(G, λ) in terms of P(H, λ) and P(K, λ), where each of H and K has either fewer vertices than G or the same number of vertices but fewer edges. Perhaps the null graph N seems to be an exception to this rule. But we find that P(N, λ) = 1 by putting K = N in III and taking G = H to be a vertex-graph.

We deduce that P(G, λ) can be found for any graph G by repeated application of the four Axioms. Perhaps the entire theory of the function P(G, λ) should be based on them. There are indeed other theorems and we shall use some of them as short cuts. For example we have Proposition 1.2, and Proposition 1.3 in the case $n > 0$. But all these can be deduced from the four Axioms, and this deduction is recommended as an exercise. Even the obvious rule that when two distinct links of G have the same two ends one of them can be deleted without altering P(G, λ) can be exhibited as a consequence of I and IV.

When P(G, λ) is defined by the Axioms it is not even necessary to require that λ shall be an integer. Any real or complex number will do. It should be noted however that the interpretation of P(G, λ) as the number of λ-colourings of G gives us an assurance that the axioms are consistent when λ is a positive integer. We need a different assurance of consistency in the more general case.

2. The graph of $P(G, \lambda)$.

Some very well-known properties of $P(G, \lambda)$ can now be dealt with in a single theorem. We write $\alpha_0(G)$ for the number of vertices of G, $\alpha_1(G)$ for the number of edges, and $p_0(G)$ for the number of components.

2.1. For any loopless graph G we can express $P(G, \lambda)$ as a polynomial in λ with integral coefficients, in such a way that the following propositions hold.

(i) The coefficient a_j of λ^j is non-zero if and only if $p_0(G) \leq j \leq \alpha_0(G)$.

(ii) $a_j = 1$ when $j = \alpha_0(G)$.

(iii) The non-zero coefficients a_j alternate in sign.

Proof.

We base this proof entirely upon the four Axioms. We have already deduced from them that $P(N, \lambda) = 1$, where N is the null graph. For any other edgeless graph G we have

$$P(G, \lambda) = \lambda^{\alpha_0(G)},$$

by repeated application of II and III. Thus the theorem holds for all edgeless graphs.

We now assume as an inductive hypothesis that the

theorem is true whenever $\alpha_1(G)$ is less than some positive integer q, and we consider the case $\alpha_1(G) = q$. We write also $\alpha_0(G) = k$ and $p_0(G) = p$.

Choose an edge A of G. By elementary graph theory $p_0(G'_A)$ is $p_0(G)$ or $p_0(G) + 1$. The latter case arises if and only if A is an isthmus of G, that is has its ends in different components of G'_A. We have in all cases $p_0(G''_A) = p_0(G)$. The contracted graph G''_A may conceivably have a loop, but only when the ends of A are joined in G by a second edge. In that case A is not an isthmus of G.

Since G'_A has fewer edges than G we can write

$$(1) \qquad P(G'_A, \lambda) = \sum_{j=p}^{k} (-1)^{k-j} n'_j \lambda^j$$

where the n'_j are positive integers, except that $n'_p = 0$ when A is an isthmus of G, and where $n'_k = 1$.

If G''_A has a loop then A is not an isthmus of G. We have also $P(G'_A, \lambda) = P(G, \lambda)$, by I and IV. Hence, by (1), the theorem holds for G.

In the remaining case, remembering that G''_A has one edge fewer, and one vertex fewer, than G, we can write

$$(2) \qquad P(G''_A, \lambda) = \sum_{j=p}^{k-1} (-1)^{k-j-1} n''_j \lambda^j ,$$

where the n''_j are positive integers. Hence, by IV,

$$(3) \qquad P(G, \lambda) = \sum_{j=p}^{k} (-1)^{k-j}(n'_j + n''_j)\lambda^j ,$$

where $n''_k = 0$. Thus the theorem is true for G, even if $n'_p = 0$.

The theorem now follows by induction.

R. C. Read has conjectured that for increasing j the non-zero numbers a_j first increase steadily to a maximum in absolute value, and then steadily diminish. No exceptions are known, but the conjecture is still unproved. ([5]).

The expression of $P(G, \lambda)$ as a polynomial in λ, for a given G, is uniquely determined. For if two polynomials agree at infinitely many values of the variable, in this case all the positive integers, they must be identical. We call this polynomial the <u>chromatic polynomial</u> or <u>chromial</u> of G. We now have the assurance of consistency called for at the end of Section 1. The Axioms are satisfied for all positive integral λ and therefore they must be satisfied by the polynomials $P(G, \lambda)$. Accordingly they remain satisfied when each $P(G, \lambda)$ is replaced by its value at any fixed real or complex value of λ.

Let us consider the behaviour of $P(G, \lambda)$ as a function of a real variable λ. As a consequence of (2.1) we have

(2.2) <u>If G is loopless and</u> $\lambda < 0$, <u>then</u> $P(G, \lambda)$ <u>is non-zero, with the sign of</u> $(-1)^{\alpha_0(G)}$.

If G is non-null we have of course $P(G, 0) = 0$. We can indeed write

$$P(G, \lambda) = \lambda Q(G, \lambda),$$

where $Q(G, \lambda)$ is another polynomial in λ. If G is connected we can obtain information about $Q(G, \lambda)$ in the region $\lambda < 1$ in the following way.

(2.3) <u>If G is non-null, loopless and connected, and if</u> $\lambda < 1$, <u>then</u> $Q(G, \lambda)$ <u>is non-zero, with the sign of</u>

$$(-1)^{\alpha_0(G) - 1}.$$

<u>Proof</u>.

If $\alpha_1(G) = 0$ then G is a vertex-graph, and $Q(G, \lambda) = 1$, by II. The theorem holds.

Assume the theorem true when $\alpha_1(G)$ is less than some positive integer q and consider the case $\alpha_1(G) = q$. Choose an edge A, with ends x and y say.

Suppose first that A is an isthmus. Then G'_A has two components H and K. Let L be the graph consisting only of A and its two ends. Then $P(L, \lambda) = \lambda(\lambda - 1)$, by (1.2). By two applications of (1.3), with n = 1, we find

$$P(G, \lambda) = \lambda^{-2} P(H, \lambda) P(K, \lambda) P(L, \lambda)$$

if $\lambda \neq 0$. We deduce that

$$Q(G, \lambda) = (\lambda - 1) Q(H, \lambda) Q(K, \lambda),$$

for all λ. (See Figure I).

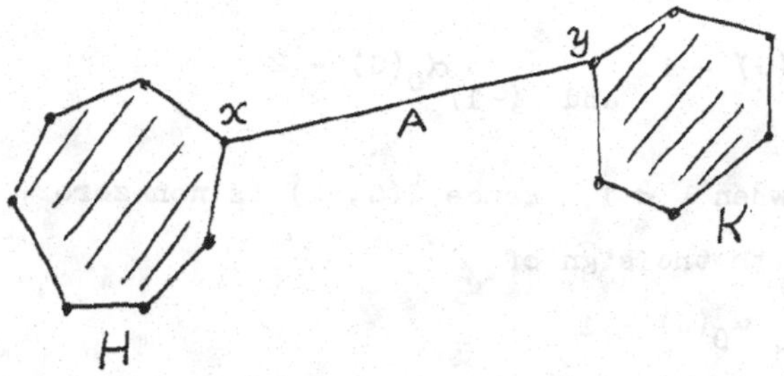

Figure I.

By the inductive hypothesis the theorem holds for H and K. We deduce that $Q(G, \lambda)$ is non-zero when $\lambda < 1$, and then has the sign of

$$(-1) . (-1)^{\alpha_0(H) + \alpha_0(K)} = (-1)^{\alpha_0(G) - 1} .$$

The theorem holds.

In the remaining case A is not an isthmus. Both

G'_A and G''_A are connected and (if loopless) satisfy the theorem. By IV,

$$Q(G, \lambda) = Q(G'_A, \lambda) - Q(G''_A, \lambda).$$

If G''_A has a loop then $Q(G''_A, \lambda) = 0$, $Q(G, \lambda) = Q(G'_A, \lambda)$, and the theorem holds for G. In the remaining case $Q(G'_A, \lambda)$ and $Q(G''_A, \lambda)$ are both non-zero, with the signs of

$$(-1)^{\alpha_0(G) - 1} \quad \text{and} \quad (-1)^{\alpha_0(G) - 2}$$

respectively when $\lambda < 1$. Hence $Q(G, \lambda)$ is non-zero when $\lambda < 1$, with the sign of

$$(-1)^{\alpha_0(G) - 1}.$$

This completes the inductive proof.

From (2.3) we deduce that

$$(d/d\lambda)P(G, \lambda)$$

is non-zero at $\lambda = 0$, and that $P(G, \lambda)$ is non-zero for $0 < \lambda < 1$ if G is non-null, loopless and connected. If G has at least one edge we have also $P(G, 1) = 0$.

Consider a graph G with an even number of vertices. Then we can sketch the graph of the chromial of the graph as in Figure II.

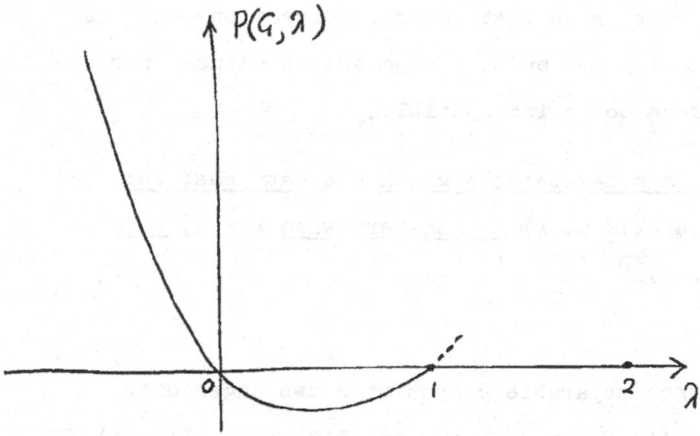

Figure II.

More information about the point $\lambda = 1$ can be obtained when G is non-separable. We recall that a graph is called <u>separable</u> if either it is disconnected or it is the union of two subgraphs H and K, each with at least one edge, such that their intersection is a vertex-graph. For example a graph with a loop or isthmus, and at least one other edge, is necessarily separable.

It can be shown that a non-separable graph G with two or more edges has the following property: for each edge A at least one of the graphs G'_A and G''_A is non-separable. In proving this we argue that if G'_A is separable G must consist of two subgraphs H and K whose intersection is a vertex graph, together with the edge A joining a vertex of H not in K to a vertex of K not in H. Next we argue that if G''_A is separable G must be the union of two subgraphs U and V, each with at least two edges, such that their intersection consists solely of A and its two ends. These two structures for G are easily seen to be incompatible.

2.4. Let G be a non-separable graph with at least two edges. Then $(d/d\lambda)P(G, \lambda)$ is non-zero when $\lambda = 1$, with the sign of $(-1)^{\alpha_0(G)}$.

Proof.

The only non-separable graphs with two edges only are the 2-circuits. For such a graph $P(G, \lambda) = \lambda(\lambda - 1)$, and the theorem holds.

Assume the theorem true when $\alpha_1(G)$ is less than some integer $q \geq 3$, and consider the case $\alpha_1(G) = q$. Choose an edge A.

By IV we have

$$(4) \quad (d/d\lambda)P(G, \lambda) = (d/d\lambda)P(G'_A, \lambda) - (d/d\lambda)P(G''_A, \lambda).$$

If G''_A has a loop it must be G'_A that is non-separable. Then, by I, the truth of the theorem for G follows from its truth for G'_A.

For the remaining case we observe that the chromial of any connected separable graph J is of the form

$$\lambda^{-1}P(H, \lambda)P(K, \lambda),$$

where H and K have at least one edge each, by (1.3).

Since H and K have edges their chromials divide by $\lambda - 1$. Hence $P(J, \lambda)$ divides by $(\lambda - 1)^2$, and its derivative vanishes when $\lambda = 1$.

Using the inductive hypothesis we can now assert that if $(d/d\lambda)P(G'_A, \lambda)$ is non-zero at $\lambda = 1$ it has the sign of $(-1)^{\alpha_0(G)}$ there. Similarly if $(d/d\lambda)P(G''_A, \lambda)$ is non-zero at $\lambda = 1$ it has the opposite sign. But at least one of these derivatives is non-zero at $\lambda = 1$ since at least one of G'_A and G''_A is non-separable. Hence, by (4), the theorem holds for G. It follows in general by induction.

It might be supposed that $P(G, \lambda)$ is non-zero throughout the interval $1 < \lambda < 2$, at least for loopless non-separable graphs. But counter-examples are easily constructed. Thus we have the graph $K_{2,3}$ with 5 vertices a_1, a_2, b_1, b_2, b_3 and 6 edges, each a_i being joined to each b_j by a single edge. (Figure III).

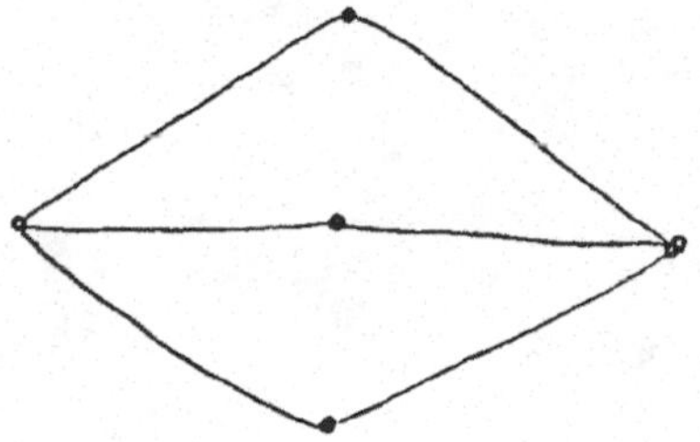

Figure III.

The graph $G = K_{2,3}$ is clearly non-separable. For values of λ in the interval $1 < \lambda < 2$ and sufficiently near 1 we have $P(G, \lambda) < 0$, by (2.4). But G is 2-colourable, and therefore $P(G, 2) > 0$. By continuity $P(G, \lambda)$ must take the value zero somewhere in the interval $1 < \lambda < 2$.

3. Planar triangulations.

Consider a connected graph G embedded in the 2-sphere so as to form a map whose faces are all triangular. This means that each face is bounded by a simple closed curve, the union of three links of G. It is impossible that G should have a loop or an isthmus, but double joins are possible. We call such a map a planar or spherical triangulation. Diagrams of some planar triangulations are shown in Figure IV. We can suppose them obtained by projecting a spherical diagram from the North Pole onto a tangent plane at the South Pole. The face containing the North Pole becomes the outer face in the plane.

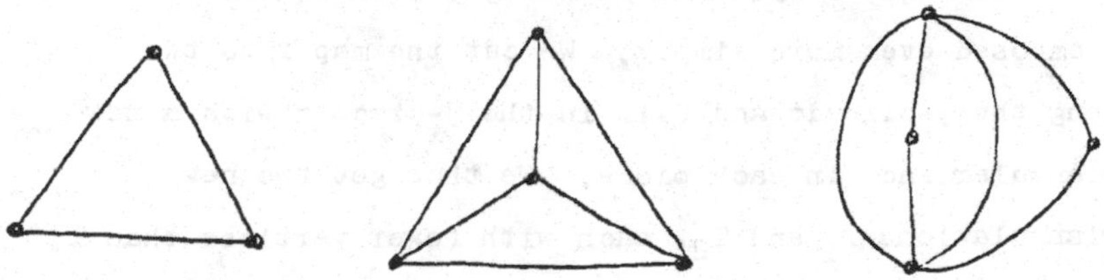

Figure IV

A separating circuit in a triangulation T is a circuit in T separating one vertex from another, that is a circuit not bounding a face. If T has a separating 2-ciruit, as in the last diagram of Figure IV, we can cut along the 2-circuit so as to separate the sphere into two parts, each bounded by a copy of the 2-circuit. In each part we fill in the 2-circuit with a new 2-sided face. We thus get two new spherical maps M_1 and M_2. In each of these we can erase one edge of the 2-circuit, fusing the 2-sided face with one of its neighbouring triangles. We thus derive triangulations T_1 and T_2 from M_1 and M_2 respectively. They each have fewer vertices than T and, by Proposition (1.3) and Axioms I and IV, we have

$$3.1 \qquad P(T, \lambda) = \frac{P(T_1, \lambda)P(T_2, \lambda)}{\lambda(\lambda - 1)} .$$

A triangle T with a separating 3-circuit can be decomposed even more simply. We cut the map into two along the 3-circuit and fill in the 3-circuit with a new triangular face in each piece. We thus get two new triangulations T_1 and T_2, each with fewer vertices than T. Applying Proposition (1.3) we find

$$3.2 \qquad P(T, \lambda) = \frac{P(T_1, \lambda)P(T_2, \lambda)}{\lambda(\lambda - 1)(\lambda - 2)} .$$

Because of these decompositions cataloguers of

chromials of triangulations usually list only "simple" triangulations, in which no separating 2-circuit or 3-circuit exists. Such catalogues have been given by Birkhoff and Lewis in ([2]), and by Ruth Bari in ([1]). There is an unpublished list of some 900 calculated by D. W. Hall and his collaborators. They represent intermediate stages in work culminating in a paper dealing with the chromial of the (dual of the) truncated icosahedron ([4]). Actually these papers deal with triangulations in their dual form as trivalent planar maps. The chromials, as in the case of the truncated icosahedron, are based on face-colourings. Moreover in these catalogues a chromial $P(T, \lambda)$ is divided by

$$\lambda(\lambda - 1)(\lambda - 2)(\lambda - 3),$$

normally a factor, before being recorded, and it is then expressed in terms of $u = \lambda - 3$ instead of λ. Division by the above product introduces a term u^{-1} for Eulerian or 3-colourable triangulations, but it is rarely necessary to consider these. Expressing the chromials in terms of u makes the numerical coefficients much simpler, though apparently no theoretical reason has as yet been given for this phenomenon.

D. W. Hall, J. W. Siry and B. R. Vanderslice gave all the zeros of the chromial (for face-colourings) of the truncated icosahedron, and the zeros of many other chromials of triangulations have been computed at Waterloo. In these investigations no real zeros have been found in the intervals $1 < \lambda < 2$ and $4 \leq \lambda$. On the other hand there always seems to be a real zero near

$$1 + \tau = \frac{3 + \sqrt{5}}{2},$$

and there are often others in $3 < \lambda < 4$.

The chromial of a triangulation has specially interesting properties at $\lambda = 1 + \tau$. It can be shown for example that

$$0 < |P(T, 1 + \tau)| \leq \tau^{5-k},$$

where k is the number of vertices of T. ([6]). The first inequality of the formula is related to the fact that $P(T, \lambda)$ must be non-zero at $\lambda = (3 - \sqrt{5})/2$, by (2.3).

Another interesting property is the "Golden Identity", relating $P(T, 1 + \tau)$ and $P(T, 2 + \tau)$. It is

$$P(T, 2 + \tau) = (2 + \tau)\,\tau^{3k-10} P^2(T, 1 + \tau).$$

(See [7]).

Attempts have been made to find other values of λ at which the properties of $P(T, \lambda)$ are unusually simple. There are the obvious values 0, 1 and 2, at which $P(T, \lambda)$ vanishes, and there is $\lambda = 3$ for which $P(T, \lambda)$ is non-zero only in the Eulerian case. Inspection of a catalogue shows that many chromials of simple triangulations divide by powers of $\lambda - 3 = u$. This is a remarkable property of the number 3, for hardly any other factorizations of chromials of simple triangulations (into polynomials with integral coefficients) are known. I write here of chromials in "reduced form", that is after division by $\lambda(\lambda - 1)(\lambda - 2)(\lambda - 3)$. However two of Dick Wick Hall's 900 have been found to divide by $u^2 - u + 1$. Their reduced chromials are

$$u^{10} + 10u^8 - 14u^7 + 47u^6 - 61u^5 + 59u^4 - 23u^3 -2u^2 + 5u - 1,$$

$$u^{12} + 12u^{10} - 15u^9 + 62u^8 - 81u^7 + 131u^6 - 93u^5 + 63u^4 - 6u^3 - 3u^2 + 4u - 1.$$

The corresponding triangulations are shown in Figures V and VI respectively.

As for other special values of λ we have of course $1 + \tau$, and the Golden Identity suggests $2 + \tau$. In the work of D. W. Hall and D. C. Lewis on the Six-ring the polynomial $\lambda^3 - 5\lambda^2 + 6\lambda - 1$ is prominent, and one of its zeros is close to a chromatic zero for the truncated icosahedron. ([3], [4]). S. Beraha has pointed out that all these special values of λ fit the formula

$$u = 2\cos(2\pi/n) = B_n$$

for integers $n \geq 2$. Thus $B_2 = 0$, $B_3 = 1$, $B_4 = 2$, $B_5 = 1 + \tau$, $B_6 = 3$ and B_7 is a root of $\lambda^3 - 5\lambda^2 + 6\lambda - 1$ (diminished by 2). Moreover $B_{10} = 2 + \tau$.

I have encountered the numbers B_n in some researches on "Chromatic Sums of Planar Triangulations" In these investigations each triangulation is rooted, that is a face, an edge and a vertex, mutually incident, are distinguished as the root-face, root-edge and root-vertex respectively. Two variables y and z are introduced and a generating function $l(y, z)$ is defined by

$$l(y, z) = \lambda(\lambda - 1)y + \sum_T y^{n(T)} z^{t(T)} P(T, \lambda),$$

where $n(T)$ is the valency of the root-vertex of the general rooted triangulation T, and $t(T)$ is the number of its faces. The object of the study was to find equations for $l(y, z)$ enabling the coefficients to be determined by an explicit general formula.

The problem reduces to triviality for the cases $\lambda = 0, 1, 2$. In the cases $\lambda = 1 + \tau$ and $\lambda = 3$ an equation for $l(y, z)$ was found and the coefficients in $h(z) = l(1, z)$ were determined. The results are to be published in the Canadian Journal of Mathematics. I now have an equation for $l(y, z)$ in the case $\lambda = B_7$, but so far I have been unable to solve it. Such success as has been achieved is related to values of λ of the form B_n.

It seems that there are now two outstanding problems in the theory of chromials of triangulations: when do chromials factorize, and what is the significance of the Beraha numbers? Oh, yes, there is the popular problem of whether a chromial can have a zero at $\lambda = 4$, but that is only a special case of the problem of factorization.

REFERENCES

[1] Ruth. A. Bari, Regular Major Maps of at most 19 Regions, and their Q-chromials, J. Combinatorial Theory, 12 (1972), 132-142.

[2] G. D. Birkhoff and D. C. Lewis, Chromatic Polynomials, Trans. Amer. Math. Soc. 60 (1946), 355-451.

[3] D. W. Hall and D. C. Lewis, Coloring Six-rings, Trans. Amer. Math. Soc., 64 (1948), 184-191.

[4] D. W. Hall, J. W. Siry and B. R. Vanderslice, The Chromatic Polynomial of the Truncated Icosahedron. Proc. Amer. Math. Soc., 16 (1965), 620-628.

[5] Ronald C. Read, An Introduction to Chromatic Polynomials, J. Combinatorial Theory, 4 (1968), 52-71.

[6] W. T. Tutte, On Chromatic Polynomials and the Golden Ratio, J. Combinatorial Theory, 9 (1970), 289-296.

[7] W. T. Tutte, The Golden Ratio in the Theory of Chromatic Polynomials, Annals of the New York Academy of Sciences, Vol. 175, Article 1, 391-402.

SOME PARTITIONS OF ALL TRIPLES INTO STEINER TRIPLE SYSTEMS

Richard M. Wilson, The Ohio State University

1. Introduction

Let t,k,v be integers with $0 \le t \le k \le v$. A Steiner system $S(t,k,v)$ is a hypergraph $(X,\mathcal{A})$ where X is a v-set, $\mathcal{A}$ is a class of k-subsets of X (called blocks), and such that for each t-subset $T \subseteq X$, there is a unique block $A \in \mathcal{A}$ with $T \subseteq A$. An $S(0,k,v)$ consists of a single k-subset of a v-set; an $S(1,k,v)$ is a partition of a v-set into k-subsets; and an $S(k,k,v)$ necessarily has blocks $\mathcal{A} = \mathcal{P}_k(X)$, the class of all k-subsets of X (i.e., an $S(k,k,v)$ is a complete k-uniform hypergraph).

If $(X,\mathcal{A})$ is an $S(t,k,v)$ and I is an i-subset of X with $0 \le i \le t$, then as is well known, the number of blocks $A \in \mathcal{A}$ such that $I \subseteq A$ is

$$b_i = \binom{v-i}{t-i} / \binom{k-i}{t-i} .$$

The existence of an $S(t,k,v)$ implies, of course, that these numbers $b_0, b_1, \cdots, b_t$ are integers. We observe that

$$b_0 = b_i \cdot \binom{v}{i} / \binom{k}{i} .$$

Thus: If Steiner systems $S(t,k,v)$ and $S(i,k,v)$ exist $(0 \le i \le t)$, then the number $\binom{v}{i}/\binom{k}{i}$ of blocks of the $S(i,k,v)$ always divides the number of blocks of the $S(t,k,v)$. It is not inconceivable, then, that there exists an $S(t,k,v)$ $(X,\mathcal{A})$ whose class of blocks admits a partition

$$\mathcal{A} = \mathcal{A}_1 \dot\cup \mathcal{A}_2 \dot\cup \cdots \dot\cup \mathcal{A}_m , \; m = \binom{v-i}{t-i} / \binom{k-i}{t-i} ,$$

such that each partial system $(X,\mathcal{A}_j)$, $j=1,2,\cdots,m$, is an $S(i,k,v)$. We call an $S(t,k,v)$ i-resolvable if it has the above property. Every $S(t,k,v)$ is trivially 0-resolvable and t-resolvable.

1-resolvable $S(2,k,v)$'s are known to exist for all positive integers v satisfying $v \equiv k \pmod{k(k-1)}$ when $k = 2,3,4$ [3] and for all sufficiently large integers v satisfying this congruence when $k > 4$ [7]. The planes of finite affine spaces over the field of two elements furnish examples of 1-resolvable $S(3,4,2^n)$'s . It is quite likely that 1-resolvable $S(3,4,v)$'s (Steiner quadruple systems) exist for all positive integers $v \equiv 4$ or $8 \pmod{12}$

It is interesting to inquire as to the i-resolvability of the $S(k,k,v)$'s . In an alternate formulation, the $S(k,k,v)$ is i-resolvable if and only if the k-subsets of a v-set can be colored in $m = \binom{v-i}{k-i}$ colors so that for each i-subset, the m k-subsets containing it receive different colors.

It has recently been shown by Zs. Baranyai [verbal communication from L. Lovász] that the $S(k,k,v)$ is 1-resolvable whenever k divides v . Nothing seems to be known concerning the 2-resolvability of the $S(k,k,v)$ except in the case $k = 3$ discussed below. The $S(4,4,10)$, $S(5,5,11)$, $S(6,6,12)$ are known not to be, respectively, 3,4,5-resolvable [6].

An $S(2,3,v)$ is called a Steiner triple system. Kirkman [4] in 1847 proved that $v \equiv 1$ or $3 \pmod 6$ is a necessary and sufficient condition for the existence of an $S(2,3,v)$. $v-2$ such systems would be required for a partition of the 3-subsets of a v-set.

In 1850, Cayley [1] observed that the $S(3,3,7)$ is not 2-resolvable and Kirkman [5] showed that the $S(3,3,9)$ is 2-resolvable. J. Doyen [2] has given lower bounds for the maximum number of block-disjoint Steiner triple systems which may be found in the $S(3,3,v)$. R.H.F. Denniston [written communication, Nov., 1972] has succeeded in showing the 2-resolvability of the $S(3,3,v)$ for v=9,13,15,19,21,25,31,33,49, and 51; and L. Teirlinck [written communication from J. Doyen, Oct., 1972] has proved that the 2-resolvability of the $S(3,3,v)$ implies the 2-resolvability of the $S(3,3,3v)$. Recently, Denniston [written communication from J. Doyen, May, 1973] has shown that the 3-subsets of a 15-set can be partitioned into 1-resolvable $S(2,3,15)$'s (Kirkman designs).

It seems quite likely that the $S(3,3,v)$ is 2-resolvable whenever

$v \equiv 1$ or $3 \pmod 6$ and $v \geq 9$. We give some further evidence for this conjecture:

<u>Theorem. The $S(3,3,v)$ is 2-resolvable if for every prime divisor p of $v-2$, the order of -2 modulo p is congruent to 2 modulo 4</u>.

We remark that $p \equiv 7 \pmod 8$ is a sufficient condition that the order of -2 modulo p be $\equiv 2 \pmod 4$, since quadratic reciprocity shows that -2 is a nonresidue and hence the order of -2 is even in addition to dividing $p-1$.

The first several values of v for which the theorem applies are $v=9,25,33,49,51,73,75,81,91,105,129,153,163,169,193,201$.

2. Proof of the Theorem

Let n be a positive integer not divisible by 2 or 3 and let G be an abelian group of order n, written additively. Let $\mathfrak{B}_o$ be the class of all subsets $A = \{x,y,z\}$ of three distinct elements of G such that $x+y+z = 0$ in G. Now a pair $\{x,y\}$ of distinct elements of G will be contained in some member A of $\mathfrak{B}_o$ (and then a unique member A) if and only if the element z determined by $x+y+z = 0$ is distinct from x and y; that is, $\{x,y\}$ is an <u>uncovered</u> pair (i.e. not contained in a member of $\mathfrak{B}_o$) if and only if $2x+y = 0$ or $x+2y = 0$ in G. Hence, by our constraints on n, $x = 0$ occurs with every element $y \neq 0$ in a member of $\mathfrak{B}_o$; but a nonzero group element x occurs in exactly two uncovered pairs, namely $\{-\frac{1}{2}x,x\}$ and $\{x,-2x\}$. Thus the graph Γ whose vertices are the $n-1$ nonzero elements of G and whose edges are the uncovered pairs is the union of disjoint polygons. These graphs Γ are shown in Fig. 1 below for the cases when G is the group Z_n of residue classes modulo n, $n=7,11,13,17$.

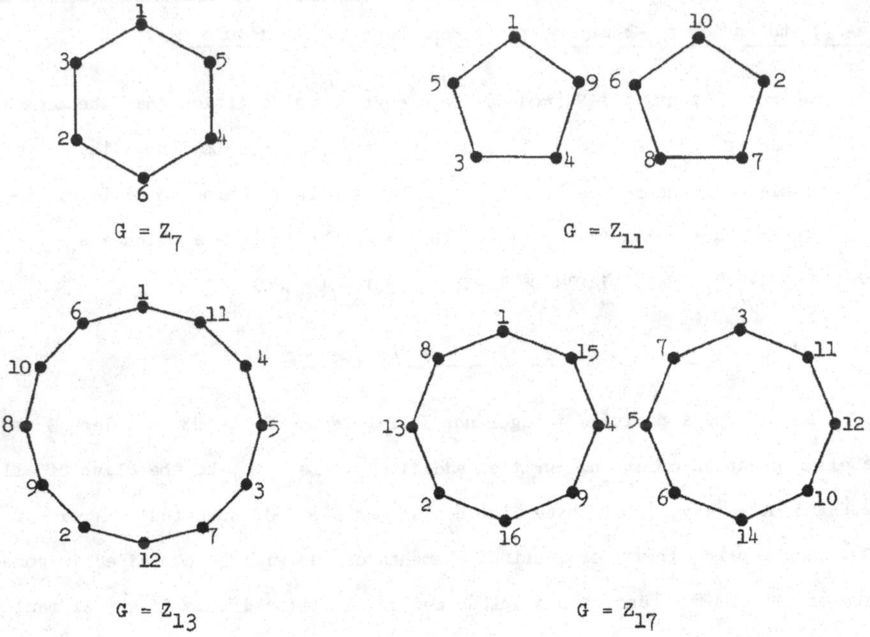

Figure 1.

Each component of the graph Γ corresponds to a cycle in the cycle decomposition of the permutation $x \mapsto -2x$ (or its inverse $x \mapsto \frac{1}{-2}x$) of the nonzero elements of G.

The length of the polygon component P_x of Γ which contains a given nonzero $x \in G$ is the least positive integer ℓ such that $(-2)^{\ell}x = x$ in G, or equivalently, $(-2)^{\ell} \equiv 1 \pmod{m}$ where m is the order of x in G. That is, the length of P_x is the order of -2 modulo m. If $m = p_1^{\alpha_1} p_2^{\alpha_2} \cdots p_r^{\alpha_r}$ is the factorization of m into powers of distinct primes, then elementary considerations show that the order of -2 modulo m is the least common multiple of the orders of -2 modulo $p_i^{\alpha_i}$, $i=1,2,\ldots,r$, and the order of -2 modulo $p_i^{\alpha_i}$ is always a nonnegative power of p_i times the order of -2 modulo p_i. Thus we have

Proposition 1. All polygons of Γ have even length if and only if the order of -2 modulo p is even for all prime divisors p of n. All polygons of Γ have length $\equiv 2 \pmod 4$ if and only if the order of -2 modulo p is $\equiv 2 \pmod 4$ for all prime divisors p of n.

Suppose now that all polygons of Γ have even length. Then Γ is bipartite (2-chromatic) and it is possible to partition the nonzero elements of G into disjoint sets S and T such that $x \in S$ implies $-2x \in T$ and $x \in T$ implies $-2x \in S$. We can then construct n Steiner triple systems of order $v = n+2$ as follows: Introduce two new points ∞, ∞' and let $X = G \dot{\cup} \{\infty,\infty'\}$. Let $\mathcal{A}_0$ be the union of $\mathcal{B}_0$, $\{\{0,\infty,\infty'\}\}$,

$$\mathcal{S}_0 = \{\{\infty,x,-2x\} : x \in S\}, \text{ and}$$
$$\mathcal{T}_0 = \{\{\infty',y,-2y\} : y \in T\}.$$

Let $\mathcal{A}_g$, $\mathcal{B}_g$, $\mathcal{S}_g$, $\mathcal{T}_g$ denote, respectively, the images of $\mathcal{A}_0$, $\mathcal{B}_0$, $\mathcal{S}_0$, $\mathcal{T}_0$ under the permutation of X which fixes ∞,∞' and takes $x \mapsto x+g$ for $x \in G$.

It is readily checked that we have

Proposition 2. Suppose Γ is bipartite and let S,T be as above. Then each system $(X,\mathcal{A}_g)$, $g \in G$, is an $S(2,3,v)$.

For example, when $G = Z_7$, $S = \{1,2,4\}$, $T = \{3,5,6\}$, the blocks of

$\mathcal{A}_0$ are

$$\{0,\infty,\infty'\} ,$$
$$\{0,1,6\} , \{0,2,5\} , \{0,3,4\} , \{1,2,4\} , \{3,5,6\} ,$$
$$\{\infty,1,5\} , \{\infty,2,3\} , \{\infty,4,6\} ,$$
$$\{\infty',3,1\} , \{\infty',6,2\} , \{\infty',5,4\} .$$

A total of seven $S(2,3,9)$'s are obtained by "developing" this basic one modulo 7 . In this case, they turn out to be disjoint and partition $P_3(X)$.

We observe that a 3-subset $\{x,y,z\}$ of G occurs in exactly one of the classes $\mathcal{B}_g$; indeed, $\mathcal{B}_g$ is the class of 3-subsets $\{x,y,z\}$ such that $x+y+z = 3g$ in G . Also, a 3-subset $\{g,\infty,\infty'\}$ of X occurs only in the class $\mathcal{A}_g$. The conditions under which the classes $\mathcal{A}_g$, $g \in G$, partition $P_3(X)$ are given in the following proposition, which together with Props. 1 and 2 completes the proof of the Theorem.

Proposition 3. Suppose Γ is bipartite and let S,T be as above. Then the following are equivalent:

(i) $\{\mathcal{A}_g : g \in G\}$ is a partition of the 3-subsets of X .

(ii) $x \in S$ if and only if $-x \in T$.

(iii) Every polygon of Γ has length congruent to 2 modulo 4 .

Proof. In discussing condition (i), it remains only to consider the 3-subsets of the form $\{\infty ,a,b\}$ and $\{\infty',a,b\}$. Now the number of $g \in G$ such that $\{\infty,a,b\} \in \mathcal{B}_g$ is equal to the number of $g \in G$ such that $\{\infty, a-g, b-g\} \in \mathcal{B}_0$, which in turn is equal to the number of 3-subsets $\{\infty,y,z\} \in \mathcal{B}_0$ such that $y-z = a-b$. Thus the number of $g \in G$ such that $\{\infty,a,b\} \in \mathcal{B}_g$ is equal to the number of times $a-b$ occurs among the differences $(x-(-2x) , (-2x)-x : x \in S) = (\pm 3x : x \in S)$. However, condition (ii) is equivalent to the assertion that each nonzero element of G occurs exactly once among $(\pm 3x: x \in S)$. The discussion of the 3-subsets $\{\infty', a, b\}$ is completely analogous, and it is now clear that (i) is equivalent to (ii).

Suppose Γ is bipartite, let $x \in G - \{0\}$, and assume that the polygon P_x of Γ containing x has length $\ell \equiv 2 \pmod 4$. So $(-2)^{\ell} \equiv 1 \pmod m$ where m is the order of x in G. Then $(-2)^{\ell} \equiv 1 \pmod{p^{\alpha}}$ for primes p where $p^{\alpha} | m$ and hence $(-2)^{\frac{1}{2}\ell} \equiv 1$ or $-1 \pmod{p^{\alpha}}$. But the former is impossible since $\frac{1}{2}\ell$ is odd while the order of -2 modulo p is even by Prop. 1. Thus $(-2)^{\frac{1}{2}\ell} \equiv -1 \pmod m$, or $(-2)^{\frac{1}{2}\ell} x = -x$ in G. That is, x and $-x$ are "opposite" vertices of the polygon P_x. It is now clear that (iii) implies (ii).

Conversely, let p be a prime divisor of n, let $x \in G$ have order p in G, and suppose the component P_x of Γ containing x has even length ℓ. Then $(-2)^{\frac{1}{2}\ell} x = -x$ in G. If $\frac{1}{2}\ell$ were even, x and $-x$ would necessarily both belong to S or to T. We conclude that (ii) implies (iii).

3. Construction of a single system

As in the previous section, let G be an abelian group of order $n \equiv 1$ or $-1 \pmod 6$. Let $\mathcal{C}$ denote the class of 3-subsets $A = \{x,y,z\}$ of $G - \{0\}$ such that $x+y+z = 0$ in G and let Γ' denote the graph with vertex set $G - \{0\}$ and edge set consisting of those pairs $\{x,y\}$ which are not contained in any member of $\mathcal{C}$. The edge set of Γ' is the union of the edge set of Γ (as in §2) and the $\frac{1}{2}(n-1)$ pairs $\{x,-x\}$, $0 \neq x \in G$.

The graphs Γ' are shown below in Fig. 2 for $G = Z_7, Z_{11}$.

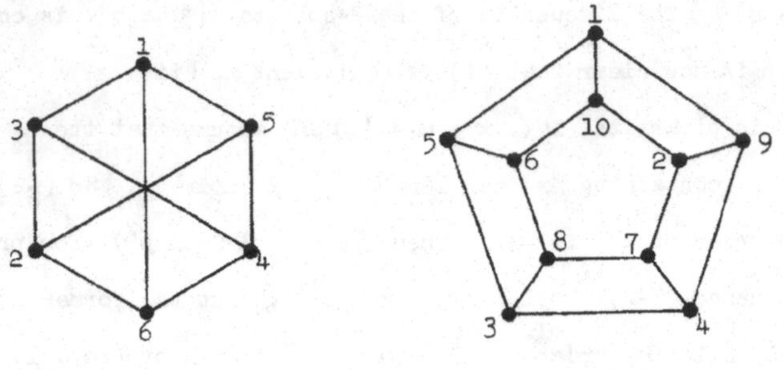

$G = Z_7$ $\qquad$ $G = Z_{11}$

Figure 2.

The component of Γ' containing a vertex x is of one of two types: (I) a polygon of even length (the component P_x of Γ) with opposite vertices joined (in the case that $(-2)^j x = -x$ for some j); or (II) two polygons (the components P_x and P_{-x} of Γ) of the same length with corresponding vertices under an isomorphism $(y \mapsto -y)$ joined.

Graphs of types (I) and (II) admit 1-factorizations, that is, it is possible to color their edges with three colors so that each vertex is incident with exactly one edge of each color (a Tait-coloring). Fig. 3 shows 1-factorizations for the graphs of Fig. 2.

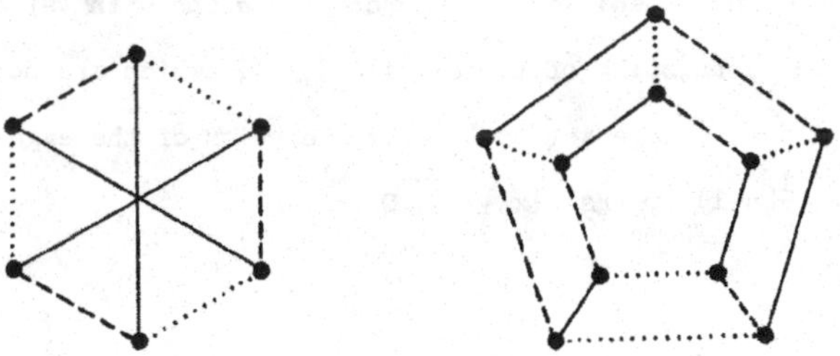

Figure 3.

Thus Γ' itself admits a 1-factorization.

Now let $X = \{\infty_1,\infty_2,\infty_3\} \cup (G-\{0\})$ and let $\mathcal{B}_i$ $(i=1,2,3)$ be the class of 3-subsets of X obtained by adding ∞_i to the edges of color i in a 1-factorization of Γ. With $\mathcal{A} = \mathcal{C} \cup \mathcal{B}_1 \cup \mathcal{B}_2 \cup \mathcal{B}_3 \cup \{\{\infty_1,\infty_2,\infty_3\}\}$, it is easily checked that $(X,\mathcal{A})$ is an $S(2,3,n+2)$.

This construction of $S(2,3,v)$'s is of note in that it is a "direct" construction which does not treat the cases $v \equiv 1$ or $3 \pmod 6$ separately.

We point out that the $S(2,3,13)$ constructed above is the "other" system of order 13. It is well known that there are exactly two nonisomorphic $S(2,3,13)$'s. The easiest to define has points Z_{13} and blocks $\mathcal{A}_1 \cup \mathcal{A}_2$ where $\mathcal{A}_1 = \{\{1+g, 3+g, 9+g\} : g \in Z_{13}\}$ and $\mathcal{A}_2 = \{\{2+g, 5+g, 6+g\} : g \in Z_{13}\}$. Now given a block $A \in \mathcal{A}$ of an $S(2,3,v)$ $(X,\mathcal{A})$, we may consider the graph Γ_A with vertices $X-A$ and edge set consisting of those pairs $\{x,y\} \subseteq X-A$ such that the block containing $\{x,y\}$ has its third point in A. For the system $(Z_{13}, \mathcal{A}_1 \cup \mathcal{A}_2)$, the graphs Γ_A are isomorphic to one of the graphs illustrated in Fig. 4 below (which are, respectively, Γ_A for $A = \{1,3,9\} \in \mathcal{A}_1$ and $A = \{2,5,6\} \in \mathcal{A}_2$). Neither is isomorphic to the right hand graph of Fig. 2.

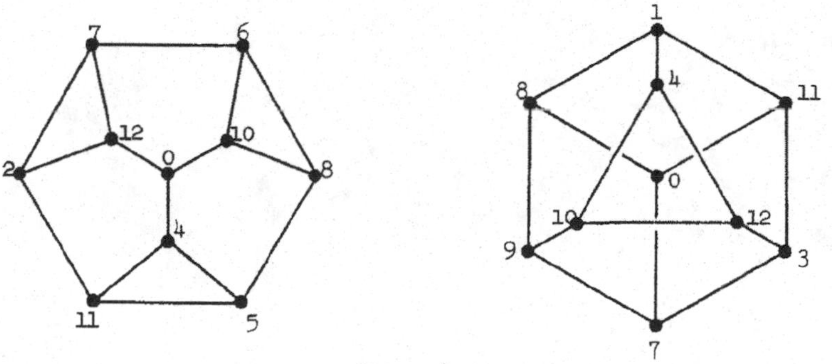

Figure 4.

As a concluding remark, we observe that if $v \equiv 5 \pmod 6$, then there exists a class $\mathcal{A}$ of subsets of a v-set X where all members of $\mathcal{A}$ have size 3 except for a single $A_o \in \mathcal{A}$ with size 5, and such that any 2-subset of X is contained in exactly one member of $\mathcal{A}$.

Let G be an abelian group of order $n = v-2 \equiv 3 \pmod 6$ which has a cyclic Sylow 3-subgroup, so that G has exactly two elements a, b of order 3. Let $\mathcal{C}$ and Γ' be defined as above. Then Γ' has one component consisting of a single edge joining a and b; all other components of Γ' are of type (I) or (II). Let $X = \{\infty_1, \infty_2, \infty_3\} \cup (G-\{0\})$, $A_o = \{a, b, \infty_1, \infty_2, \infty_3\}$, and let $\mathcal{B}_i$ $(i=1,2,3)$ be the class of 3-subsets of X obtained by adding ∞_i to the edges of color i in a 1-factorization of Γ'-a-b. Then $\mathcal{A} = \mathcal{C} \cup \mathcal{B}_1 \cup \mathcal{B}_2 \cup \mathcal{B}_3 \cup \{A_o\}$ has the above stated property.

REFERENCES

1. A. Cayley, On the triadic arrangements of seven and fifteen things, London, Edinburgh and Dublin Philos. Mag. and J. Sci. (3) 37 (1850), 50-53.

2. J. Doyen, Constructions of disjoint Steiner triple systems, to appear.

3. H. Hanani, D. K. Ray-Chaudhuri, and R. M. Wilson, On resolvable designs, Discrete Math. 3 (1972), 343-357.

4. T. P. Kirkman, On a problem in combinations, Cambridge and Dublin Math. J. 2 (1847), 191-204.

5. T. P. Kirkman, Note on an unanswered prize question, Cambridge and Dublin Math. J. 5 (1850), 255-262.

6. E. S. Kramer and D. M. Mesner, Intersections among Steiner systems, Notices of the Amer. Math. Soc. 19 (1972), page A-563 (Abstract No. 72T-A161).

7. D. K. Ray-Chaudhuri and R. M. Wilson, The existence of resolvable block designs, in: A Survey of Combinatorial Theory (J. N. Srivastava et. al., ed.), North-Holland/American Elsevier Publ. Co., Amsterdam/New York, 1973.

UNSOLVED PROBLEMS

1. Let $H = (E_i / i \in I)$ be a hypergraph with the Helly property. Given an integer t, let

$$H^t = (E_{i_1} \cup E_{i_2} \cup \ldots \cup E_{i_t} / i_1, i_2, \ldots, i_t \in I).$$

Conjecture : If any t+1 edges of H^t have a non-empty intersection, then $\tau(H^t) \leq t$.

(True for $t \leq 3$ if H is a family of subsets A_i of a tree T such that T_{A_i} is a tree).

2. Problem : Characterize the simple graphs $G = (X, \Gamma)$ such that $H_G = (\Gamma(x) \cup \{x\} / x \in X)$ is a balanced hypergraph.

(Jaeger and Payan have shown that if a graph G is without cycle of length > 3, then H_G is balanced).

3. Let $\mathcal{C} = (C_1, C_2, \ldots, C_n)$ be n convex sets in R^2.

Let $L^3(\mathcal{C})$ denote the 3-uniform hypergraph with vertices $c_1, c_2, \ldots, c_n$ representing the members of $\mathcal{C}$, $c_i c_j c_k$ being an edge in $L^3(\mathcal{C})$ if and only if $C_i \cap C_j \cap C_k \neq \emptyset$.

Problem : Characterize the hypergraphs H isomorphic to a $L^3(\mathcal{C})$. (These hypergraphs H are not necessarily balanced, but have similar properties).

4. S. Ulam has proposed the following conjecture :

If X is the set of all integer points in the plane, two disjoint sets $A \subseteq X$ and $B \subseteq X$ are called *neighbours* if $|a-b| \leq 1$, $|a'-b'| \leq 1$ for some $(a,a') \in A$, $(b,b') \in B$. Then, for any partition of X whose classes are bounded by a given integer, there exists a class that has six neighbours.

C.BERGE.

5. Given any simple hypergraph H (no repeated edges) with m vertices and m edges, there exists an (m-1)-vertex simple subhypergraph of H. (See [1]).

One can pose general problems of this type. Write :

$$(m,k) \to (n,\ell)$$

if every simple hypergraph with m vertices and k edges contains a simple partial subhypergraph with n vertices and ℓ edges. The above result, expressed in this language, is :

$$(m,m) \to (m-1,m).$$

N.Sauer [3] has shown that

$$\left(m, 1 + \sum_{i=0}^{n-1} \binom{m}{i}\right) \to (n, 2^n) ,$$

and A. Hajnal asked whether this could be strengthened to :

$$\left(m, 1 + \sum_{i=0}^{n-1} \binom{m}{i}\right) \to \left(m-1, 1 + \sum_{i=0}^{n-1} \binom{m-1}{i}\right), \quad m > n.$$

Recently, M.Lewin [2] verified Hajnal's conjecture in the case n = 2 :

$$(m,m+2) \to (m-1,m+1).$$

[1] J.A. Bondy, Induced Subsets, J. Combinatorial Theory, 12, B(1972) 201-202.

[2] M. Lewin, A note on induced subsets, preprint.

[3] N. Sauer. Unpublished.

J.A. BONDY.

6. A hypergraph $(X,\mathcal{E})$ is *m-intersecting* if :

$$E_1 \cap E_2 \cap \dots \cap E_m \neq \emptyset \text{ whenever } E_1, E_2, \dots, E_m \in \mathcal{E}.$$

An *(n,k,m)-hypergraph* is a hypergraph $H = (X,\mathcal{E})$ such that :

(i) $|X| = n$,

(ii) H is k-uniform,

(iii) every m-intersecting partial hypergraph of H is also (m+1)-intersecting.

Let $F(n,k,m)$ be the largest possible number of edges in an (n,k,m)-hypergraph. Erdös, Chao Ko and Rado proved that $F(n,k,1) = \binom{n-1}{k-1}$ whenever $n \geq 2k$. Erdös conjectured that $F(n,k,2) = \binom{n-1}{k-1}$ for $n \geq n_0(k)$ and $k \geq 3$. I proved that $F(n,k,,k-1) = \binom{n-1}{k-1}$ for $n \geq k+2$.

CONJECTURE. $F(n,k,m) = \binom{n-1}{k-1}$ whenever $1 \leq m < k$ and $n \geq 2k + 1 - m$.

7. An *independence system* is a hypergraph $(X,\mathfrak{E})$ such that $E \subset F \in \mathfrak{E}$ implies $E \in \mathfrak{E}$. A hypergraph is *intersecting* if it has no two disjoint edges.

CONJECTURE. Let H be a finite independence system with $\mathfrak{E} \neq \{\emptyset\}$. If H_0 is a partial hypergraph of H and H_0 is intersecting then H_0 has at most $\delta(H)$ edges. (I offer \$ 10.00 for a proof or disproof).

If E, F are two sets of positive integers and there is a one-to-one mapping $f : E \rightarrow F$ with $f(x) \geq x$ for all $x \in E$, then we write $E < F$. I can prove a weaker form of the conjecture, where the first sentence is replaced by "Let H be a hypergraph with $X = \{1,2,\ldots,n\}$, $\mathfrak{E} \neq \{\emptyset\}$ and such that $E < F \in \mathfrak{E}$ implies $E \in \mathfrak{E}$."

It would be interesting to prove the conjecture for independence systems whose maximal edges form the lines of a projective plane.

8. A *positional game* on a finite hypergraph H is played as follows :

Two players take turns to claim a previously unclaimed vertex of H. If a player succeeds to claim all the vertices of an edge of H, then he wins. If all the vertices of H have been already claimed but neither player has won yet, the game is a draw. An easy argument shows that the second player cannot have a winning strategy. If $\chi(H) > 2$ then the game cannot be a draw and so the first player has a winning strategy. In particular, let us consider the game on a hypergraph $W(n,k) = (X,\mathfrak{E})$ such that $X = \{1,2,\ldots,n\}$ and $E \in \mathfrak{E}$ if, and only if, $|E| = k$ and the elements of E form an arithmetic progression. Van der Waerden proved that, given any k, there is always an N with $\chi(W(N,k)) > 2$. Let us denote the smallest such N by $N(k)$; let us denote the smallest n such that the first player has a winning strategy on $W(n,k)$ by $n(k)$. Then $n(k) \leq N(k)$. The existing upper bounds on $N(k)$ are beyond the range of algebraic expressions. However, the numbers $N(k)$ seem to give rather crude upper bounds on $n(k)$:

one has

$$n(3) = 5 \ , \ n(4) = 13 \ , \ N(3) = 9 \ , \ N(4) = 35.$$

Find a decent upper bound on $n(k)$. Is $n(k)$ always odd ? If so, is the winning first move $\frac{1}{2}(n(k) + 1)$? Has the first player winning strategies for all $W(n,k)$ with $n \geq n(k)$?

9. One way to generalise the concept of a chromatic polynominal to hypergraphs is as follows. Given a hypergraph $H = (X,\mathcal{E})$, denote by $P(H,\lambda)$ the number of weak λ-colorings of H. Then, obviously,

$$P(H,\lambda) = \sum_{k=1}^{n} a_k \binom{\lambda}{k} k! = \sum_{k=1}^{n} b_k \lambda^k$$

where a_k is the number of partitions of X into k non-empty pairwise disjoints sets $X_1, X_2, \ldots, X_k$ such that no X_i contains an edge of H. The chromatic polynomials of hypergraphs lack some of the characteristic properties of the chromatic polynomials of graphs. For instance, $\lambda^4 - \lambda^3 - 2\lambda^2 + 2\lambda$ is a chromatic polynomial of a hypergraph, but its coefficients do not alternate signs. However, the following may still be true:

CONJECTURE. <u>For each hypergraph H, the numbers $a_1, a_2, \ldots, a_n$ first increase and then decrease.</u> The only existing result in this direction seems to be the inequality:

$$a_{k+1} \geq a_k \cdot \frac{2^{n/k} - 2}{k + 1}$$

This is obtained by an easy counting argument.

V. CHVÁTAL.

10. Can the complete graph on $2n$ vertices be decomposed into $2n-1$ perfect matchings such that the union of any two is a spanning cycle ?

A. EHRENFEUCHT, V. FABER
S. FAJAITLOVICH, J. MYCIELSKI.

11. It is well known that an ordinary graph of chromatic number k has at least $\binom{k}{2}$ edges, equality only for K_k. This result is not true for r-graphs e.g. for r = 3. Every three-chromatic hypergraph has 7 edges and the 7 Steiner triples show that this result is best possible. The corresponding complete 3-graph has 10 edges. Perhaps for sufficiently large chromatic number the conjecture nevertheless holds i.e. for every n-chromatic r-graph has at least $\binom{(r-1)(n-1)+1}{r}$ edges, with equality only for the complete r-graph of (r-1)(n-1)+1 vertices.

12. Prove that there is an absolute constant c so that in every 3-chromatic r-graph there is a vertex contained in $(1+c)^r$ edges.

In fact, it would be interesting to determine the largest integer f(r) so that every 3-chromatic r-graph has a vertex contained in f(r) r-tuples. f(2) = 2, f(3) = 3, f(4) is unknown.

P.ERDÖS.

13. Let G be a graph of chromatic number $\aleph_0$. Is it true that $\sum \frac{1}{r}$ diverges, where G has a circuit of size r ?

Let G(n ; m) be a graph of n vertices and m edges. Consider :

$$\min \sum \frac{1}{r} = f(n\ ;\ m)$$

where the sum is extended over all r's for which our graph has a circuit of size r and the minimum is taken over all G(n ; m). To prove the previous conjecture it would suffice to show that f(n ; m) tends to infinity as $\frac{m}{n}$ tends to infinity. Probably $f(n\ ;\ m) > c \log \frac{m}{n}$.

P.ERDÖS and A.HAJNAL.

14. Does there exist a 3-critical 3-uniform hypergraph so that every point has valency ≥ 7 ?

P.ERDÖS and L.LOVÁSZ.

15. Is there a $k > 2$ so that one can color the k-tuples of a set $|L| = 2k$ by $k+1$ colors so that for every $L_1 \subset L$, $|L_1| = k+1$, all $k+1$ colors occur among the k-subsets of L_1 ? For $k=2$ this is trivially possible.

The next doubtful case is $k=6$.

P.ERDÖS and M.ROSENFELD.

16. Let H be a given 3-chromatic r-graph any two edges of which have a non-empty intersection. Is it true that the number of vertices is $< c_1 r^2$?

Is it true that there are two edges which meet in $c_2 r$ points ? Are there other examples than the Fano geometry where there are no two edges which meet in $r-1$ points ?

P.ERDÖS and L.SHELAH.

17. Let A_k be the set of constants $\alpha \in [0,1]$ with the following properties :

(i) There exists a $\beta = \beta(\alpha) > \alpha$ such that for any sequence H^n of hypergraphs satisfying

$$(1)\ \underline{\lim}\ \frac{m(H^n)}{\binom{n}{k}} > \alpha \quad (H^n \text{ has } n \text{ vertices and } m(H^n) \text{ edges})$$

one can find a sequence of partial subhypergraphs $\overline{H}^{p_n} \subset H^n$ such that $p_n \to \infty$ and

$$(2)\ \lim \frac{m(\overline{H}^{p_n})}{\binom{p_n}{k}} > \beta.$$

(ii) If (1) is replaced by :

$$\underline{\lim}\ \frac{m(H^n)}{\binom{n}{k}} \geq \alpha$$

then (2) does not necessarily hold.

It is known that :

$$A_2 = \{1 - \frac{1}{k} : k \text{ is a positive integer}\}.$$

What is A_3 ?

(Reference : Erdös - Stone theorem).

P.ERDÖS and M.SIMONOVITS.

18. Find the best possible function, $f(\delta,r)$, such that $q(H) \leq f(\delta,r)$ for all r-uniform hypergraphs of maximum valency δ. It is easy to show that

(i) $f(\delta,r) \leq r(\delta-1) + 1, \quad \forall\ \delta, r$;

(ii) $f(\delta,r) = r(\delta-1) + 1$ if an $(r(\delta-1) + 1, \delta, 1)$ - balanced incomplete block design exists ;

(iii) $\frac{\delta}{r-1} (r^2-3r+3) \leq f(\delta,r)$ if $r-1$ divides δ.

V.FABER and L.LOVÁSZ.

19. Is $K_{7,7}$ minus a perfect matching the representative graph of a family of convex sets in the plane (no three members of the family have a common point) ?

20. Let H be a hypergraph. Suppose that for each partial hypergraph H' :

$$\nu(H')(\delta(H') + 1) \geq m(H')$$

Then is it true that :

$$q(H) \leq \delta(H) + 1 \ ?$$

Here :

ν = number of edges in a maximum matching

δ = maximum degree

q = chromatic index.

This would be a generalization of Vizing's theorem. Removing +1 after δ in the assumption as well as in the conclusion, we get a valid characterisation of normal hypergraphs.

21. Erdös proposed the problem of determining r-uniform hypergraphs which are 3-chromatic and any two edges of which intersect ("Clique"). We know that such a hypergraph has less than r^r edges, and we have the following example of

$[(e-1)r!]$ edges : Let $V(H) = \{(i,j) : 1 \leq i \leq j \leq r\}$, and let the edges of H be all sets of form $\{(1,\nu),(2,\nu),\ldots,(\nu,\nu),(a_{\nu+1},\nu+1),(a_{\nu+2},\nu+2),\ldots,(a_r,r)\}$ where $1 \leq a_j \leq j$.

Conjecture. The hypergraph defined above has the maximum number of edges among all 3-chromatic r-uniform "cliques".

22. It is well known that a Steiner triple system has chromatic number ≥ 3. The general problem is whether other symmetry conditions on a hypergraph also imply that the hypergraph cannot be 2-chromatic. As an example, consider the hypergraph $H_{n,k}$ obtained from the complete 2-graph K_n by considering the edges of K_n as vertices of $H_{n,k}$ and the edges of any k-clique in K_n as an edge of $H_{n,k}$. By Ramsey's theorem, $H_{n,k}$ has chromatic number ≥ 3 if n is large enough.

Problem: Try to find symmetry conditions on H which already imply it is $\geq$ 3-chromatic.

For example, is it true that an r-uniform hypergraph, whose automorphism group is transitive on the vertices (and/ or edges), primitive, the k-tuples of points have at most f(k) orbits (k=1,...,r-1),(f(k) is a suitable function), and which has enough points, cannot be 2-chromatic ?

L.LOVÁSZ.

23. Let X be a set of cardinality n. Let h be a positive integer $(h \leq n)$. Let $K_n^h = (X,\mathfrak{E})$ be the h-uniform complete hypergraph (the vertices are the points of X, the edges all the subsets of X with cardinality h).

A *clique* $\mathfrak{C} = (E_i | i \in I)$ of K_n^h is a subset of $\mathfrak{E}$ such that :

$$E_i \cap E_j \neq \emptyset \text{ for all } i \text{ and } j \text{ in } I.$$

Problem : What is the minimum cardinality of a maximal clique in K_n^h ?

Conjecture. Let $\omega'(K_n^h)$ = this number, $n \geq h(h-1)+1$ and $h = p^\alpha + 1$, where p is a prime number and α an integer; I conjecture that :

$$\omega'(K_n^h) = h(h-1)+1.$$

Remarks : In the case of the conjecture, I can prove that :

$$\omega'(K_n^h) \leq h(h-1)+1$$

To see this, take $\mathcal{G} \simeq P_h$, where P_h is the projective plane of order h-1 (the vertices of P_h are the points of P_h and the edges of P_h the lines of P_h). If h=3, I can prove that $\omega'(K_n^3) = 7$.

Jean-Claude MEYER.

24. Consider sets of squares in the plane having sides parallel to the x- or y-axis and corners at lattice points. Let a(S) denote the area covered by the union of such a set S. T.Rado conjectured that

$$\max_{T \subseteq S} \frac{a(T)}{a(S)} \geq \frac{1}{4}$$

if the squares in T are pairwise disjoint.

Ajtai showed that the conjecture is false in general. The conjecture is true if the lengths of sides of squares in S are all 1. It is not known whether the conjecture is true or not if the lengths of sides are 1 or 2.

R. RADO.

25. Let h,k, $h \geq 2$, $k \geq 1$, be integers. Prove that if G is a k-connected graph such that $\alpha(G) \leq h+k-1$, where $\alpha(G)$ is the stability number of G, then G has a spanning tree with $\leq h$ vertices of degree 1. This conjecture has been proved by P.Erdös and V.Chvátal (1971) for h = 2 and every k, by M.Las Vergnas (1971) for k = 2,3,4 (every h) and k = 5, h = 3. Of course this would be a consequence of the following stronger conjecture : if G is k-connected and $\alpha(G) \leq h+k-1$ then G contains a cycle of length $\geq n-h+1$, where n is the number of vertices of G, $h,k \geq 1$ (proved for h = 1 by P.Erdös and V. Chvátal).

M.LAS VERGNAS.

26. Let G_n be the complete bipartite graph $K_{n,n}$ minus a perfect matching. For every $n \geq 1$, G_n is the intersection graph of a family of arc-connected subsets of the plane R^2 ; is it true that G_7 is not the intersection graph of a family of convex subsets of R^2 ? (It can easily be seen that G_6 is the intersection graph of a family of convex subsets of R^2).

M. LAS VERGNAS

27. Let $H = (X,\mathcal{E})$, $\mathcal{E} = (E_i : i \in I)$ be a hypergraph such that $|E_i| \geq r$, $\forall i \in I$. Let $H_r = (P_r(X), \mathcal{E}^r)$, $\mathcal{E}^r = (E_i^R : i \in I)$ where $P_r(X)$ is the set of r-element subsets of X and E_i^r is the set of r-element subsets of E_i, $i \in I$. H_r will be called the r-th derivative of H. Note that $|E_i^r| = \binom{|E_i|}{r}$. Find necessary and sufficient conditions under which a hypergraph K is the r-th derivative of some hypergraph H.

28. Let X be a set of v elements and L be the lattice of subsets of X. A combinatorial design with X as the set of points can be viewed as a set of elements of L satisfying certain properties. Problems on combinatorial design can be asked on the projective spaces or any geometric lattice. Let PG(n,q) be the n-dimensional projective space with $(q+1)$ points on a line, $n \geq 3$. Let $1 \leq t \leq k \leq n$. Does there exist a set D of k-flats such that every t-flat is contained in exactly one k-flat of D ? For $n = 6$, $q = 2$, $t = 2$ and $k = 3$, the arithmetic conditions are satisfied and the problem is to find a set D of 127×3 planes such that every line is contained in exactly one plane of D. More generally one would like to find the smallest number of k-flats which cover all the t-flats.

D.K. RAY-CHAUDHURI.

Vol. 310: B. Iversen, Generic Local Structure of the Morphisms in Commutative Algebra. IV, 108 pages. 1973. DM 16,–

Vol. 311: Conference on Commutative Algebra. Edited by J. W. Brewer and E. A. Rutter. VII, 251 pages. 1973. DM 22,–

Vol. 312: Symposium on Ordinary Differential Equations. Edited by W. A. Harris, Jr. and Y. Sibuya. VIII, 204 pages. 1973. DM 22,–

Vol. 313: K. Jörgens and J. Weidmann, Spectral Properties of Hamiltonian Operators. III, 140 pages. 1973. DM 16,–

Vol. 314: M. Deuring, Lectures on the Theory of Algebraic Functions of One Variable. VI, 151 pages. 1973. DM 16,–

Vol. 315: K. Bichteler, Integration Theory (with Special Attention to Vector Measures). VI, 357 pages. 1973. DM 26,–

Vol. 316: Symposium on Non-Well-Posed Problems and Logarithmic Convexity. Edited by R. J. Knops. V, 176 pages. 1973. DM 18,–

Vol. 317: Séminaire Bourbaki – vol. 1971/72. Exposés 400–417. IV, 361 pages. 1973. DM 26,–

Vol. 318: Recent Advances in Topological Dynamics. Edited by A. Beck, VIII, 285 pages. 1973. DM 24,–

Vol. 319: Conference on Group Theory. Edited by R. W. Gatterdam and K. W. Weston. V, 188 pages. 1973. DM 18,–

Vol. 320: Modular Functions of One Variable I. Edited by W. Kuyk. V, 195 pages. 1973. DM 18,–

Vol. 321: Séminaire de Probabilités VII. Edité par P. A. Meyer. VI, 322 pages. 1973. DM 26,–

Vol. 322: Nonlinear Problems in the Physical Sciences and Biology. Edited by I. Stakgold, D. D. Joseph and D. H. Sattinger. VIII, 357 pages. 1973. DM 26,–

Vol. 323: J. L. Lions, Perturbations Singulières dans les Problèmes aux Limites et en Contrôle Optimal. XII, 645 pages. 1973. DM 42,–

Vol. 324: K. Kreith, Oscillation Theory. VI, 109 pages. 1973. DM 16,–

Vol. 325: Ch.-Ch. Chou, La Transformation de Fourier Complexe et L'Equation de Convolution. IX, 137 pages. 1973. DM 16,–

Vol. 326: A. Robert, Elliptic Curves. VIII, 264 pages. 1973. DM 22,–

Vol. 327: E. Matlis, 1-Dimensional Cohen-Macaulay Rings. XII, 157 pages. 1973. DM 18,–

Vol. 328: J. R. Büchi and D. Siefkes, The Monadic Second Order Theory of All Countable Ordinals. VI, 217 pages. 1973. DM 20,–

Vol. 329: W. Trebels, Multipliers for (C, α)-Bounded Fourier Expansions in Banach Spaces and Approximation Theory. VII, 103 pages. 1973. DM 16,–

Vol. 330: Proceedings of the Second Japan-USSR Symposium on Probability Theory. Edited by G. Maruyama and Yu. V. Prokhorov. VI, 550 pages. 1973. DM 36,–

Vol. 331: Summer School on Topological Vector Spaces. Edited by L. Waelbroeck. VI, 226 pages. 1973. DM 20,–

Vol. 332: Séminaire Pierre Lelong (Analyse) Année 1971-1972. V, 131 pages. 1973. DM 16,–

Vol. 333: Numerische, insbesondere approximationstheoretische Behandlung von Funktionalgleichungen. Herausgegeben von R. Ansorge und W. Törnig. VI, 296 Seiten. 1973. DM 24,–

Vol. 334: F. Schweiger, The Metrical Theory of Jacobi-Perron Algorithm. V, 111 pages. 1973. DM 16,–

Vol. 335: H. Huck, R. Roitzsch, U. Simon, W. Vortisch, R. Walden, B. Wegner und W. Wendland, Beweismethoden der Differentialgeometrie im Großen. IX, 159 Seiten. 1973. DM 18,–

Vol. 336: L'Analyse Harmonique dans le Domaine Complexe. Edité par E. J. Akutowicz. VIII, 169 pages. 1973. DM 18,–

Vol. 337: Cambridge Summer School in Mathematical Logic. Edited by A. R. D. Mathias and H. Rogers. IX, 660 pages. 1973. DM 42,–

Vol: 338: J. Lindenstrauss and L. Tzafriri, Classical Banach Spaces. IX, 243 pages. 1973. DM 22,–

Vol. 339: G. Kempf, F. Knudsen, D. Mumford and B. Saint-Donat, Toroidal Embeddings I. VIII, 209 pages. 1973. DM 20,–

Vol. 340: Groupes de Monodromie en Géométrie Algébrique. (SGA 7 II). Par P. Deligne et N. Katz. X, 438 pages. 1973. DM 40,–

Vol. 341: Algebraic K-Theory I, Higher K-Theories. Edited by H. Bass. XV, 335 pages. 1973. DM 26,–

Vol. 342: Algebraic K-Theory II, "Classical" Algebraic K-Theory, and Connections with Arithmetic. Edited by H. Bass. XV, 527 pages. 1973. DM 36,–

Vol. 343: Algebraic K-Theory III, Hermitian K-Theory and Geometric Applications. Edited by H. Bass. XV, 572 pages. 1973. DM 38,–

Vol. 344: A. S. Troelstra (Editor), Metamathematical Investigation of Intuitionistic Arithmetic and Analysis. XVII, 485 pages. 1973. DM 34,–

Vol. 345: Proceedings of a Conference on Operator Theory. Edited by P. A. Fillmore. VI, 228 pages. 1973. DM 20,–

Vol. 346: Fučík et al., Spectral Analysis of Nonlinear Operators. II, 287 pages. 1973. DM 26,–

Vol. 347: J. M. Boardman and R. M. Vogt, Homotopy Invariant Algebraic Structures on Topological Spaces. X, 257 pages. 1973. DM 22,–

Vol. 348: A. M. Mathai and R. K. Saxena, Generalized Hypergeometric Functions with Applications in Statistics and Physical Sciences. VII, 314 pages. 1973. DM 26,–

Vol. 349: Modular Functions of One Variable II. Edited by W. Kuyk and P. Deligne. V, 598 pages. 1973. DM 38,–

Vol. 350: Modular Functions of One Variable III. Edited by W. Kuyk and J.-P. Serre. V, 350 pages. 1973. DM 26,–

Vol. 351: H. Tachikawa, Quasi-Frobenius Rings and Generalizations. XI, 172 pages. 1973. DM 18,–

Vol. 352: J. D. Fay, Theta Functions on Riemann Surfaces. V, 137 pages. 1973. DM 16,–

Vol. 353: Proceedings of the Conference. on Orders, Group Rings and Related Topics. Organized by J. S. Hsia, M. L. Madan and T. G. Ralley. X, 224 pages. 1973. DM 20,–

Vol. 354: K. J. Devlin, Aspects of Constructibility. XII, 240 pages. 1973. DM 22,–

Vol. 355: M. Sion, A Theory of Semigroup Valued Measures. V, 140 pages. 1973. DM 16,–

Vol. 356: W. L. J. van der Kallen, Infinitesimally Central-Extensions of Chevalley Groups. VII, 147 pages. 1973. DM 16,–

Vol. 357: W. Borho, P. Gabriel und R. Rentschler, Primideale in Einhüllenden auflösbarer Lie-Algebren. V, 182 Seiten. 1973. DM 18,–

Vol. 358: F. L. Williams, Tensor Products of Principal Series Representations. VI, 132 pages. 1973. DM 16,–

Vol. 359: U. Stammbach, Homology in Group Theory. VIII, 183 pages. 1973. DM 18,–

Vol. 360: W. J. Padgett and R. L. Taylor, Laws of Large Numbers for Normed Linear Spaces and Certain Fréchet Spaces. VI, 111 pages. 1973. DM 16,–

Vol. 361: J. W. Schutz, Foundations of Special Relativity: Kinematic Axioms for Minkowski Space Time. XX, 314 pages. 1973. DM 26,–

Vol. 362: Proceedings of the Conference on Numerical Solution of Ordinary Differential Equations. Edited by D. Bettis. VIII, 490 pages. 1974. DM 34,–

Vol. 363: Conference on the Numerical Solution of Differential Equations. Edited by G. A. Watson. IX, 221 pages. 1974. DM 20,–

Vol. 364: Proceedings on Infinite Dimensional Holomorphy. Edited by T. L. Hayden and T. J. Suffridge. VII, 212 pages. 1974. DM 20,–

Vol. 365: R. P. Gilbert, Constructive Methods for Elliptic Equations. VII, 397 pages. 1974. DM 26,–

Vol. 366: R. Steinberg, Conjugacy Classes in Algebraic Groups (Notes by V. V. Deodhar). VI, 159 pages. 1974. DM 18,–

Vol. 367: K. Langmann und W. Lütkebohmert, Cousinverteilungen und Fortsetzungssätze. VI, 151 Seiten. 1974. DM 16,–

Vol. 368: R. J. Milgram, Unstable Homotopy from the Stable Point of View. V, 109 pages. 1974. DM 16,–

Vol. 369: Victoria Symposium on Nonstandard Analysis. Edited by A. Hurd and P. Loeb. XVIII, 339 pages. 1974. DM 26,–

Vol. 370: B. Mazur and W. Messing, Universal Extensions and One Dimensional Crystalline Cohomology. VII, 134 pages. 1974. DM 16,–